**Appropriate
Technology
and Social Values—
A Critical Appraisal**

Appropriate Technology and Social Values– A Critical Appraisal

edited by
Franklin A. Long
Cornell University

Alexandra Oleson
*American Academy of Arts
  and Sciences*

Published in association with the American Academy
of Arts and Sciences

Ballinger Publishing Company • Cambridge, Massachusetts
A Subsidiary of Harper & Row, Publishers, Inc.

This book is printed on recycled paper.

This volume is based upon work supported by the National Science Foundation (Grant No. OSS76–17285 AO3), the Johnson Foundation, and the Exxon Corporation. Any opinions, findings, conclusions or recommendations expressed in this publication are those of the authors and do not necessarily reflect the views of the supporting foundations.

International Standard Book Number: ISBN 0–88410–373–0

Library of Congress Catalog Card Number: 79–18528

Printed in the United States of America

Library of Congress Cataloging in Publication Data

Main entry under title:

Appropriate technology and social values.

   1. Technology—Social aspects—Addresses, essays, lectures.
2. Underdeveloped areas—Technology—Addresses, essays, lectures.
I. Long, Franklin A., 1910– II. Oleson, Alexandra, 1939–
III. American Academy of Arts and Sciences
T14.5.A67   301.24'3   79–18528
ISBN 0–88410–373–0

Contents

NA

Preface

The papers in this volume were originally presented at an international symposium held in June 1978 at the Wingspread Conference Center of the Johnson Foundation in Racine, Wisconsin. Organized by the American Academy of Arts and Sciences, the symposium was held under the auspices of the International Pugwash Council and the U.S. Pugwash Committee. The Pugwash movement, long recognized for its concern with arms control, has in recent years become actively involved with issues relating to science and development.

This study was made possible by grants from the National Science Foundation, the Johnson Foundation, and the Exxon Corporation. We acknowledge our indebtedness to these organizations for their generous support and their understanding of the importance of the topic. To William Blanpied of the National Science Foundation we owe special thanks both for his encouragement and advice and for his lively and provocative contributions at Wingspread.

In the course of planning, writing, and editing this volume, we have benefited from the advice and assistance of many friends and colleagues. Harvey Brooks, Paul De Forest, Bernard Feld, Robert Kates, T.W. Schultz, Robert Socolow, and John Voss served with us on the Academy Steering Committee. Paul De Forest played a major role in planning and organizing the symposium, and Mary Hughes provided invaluable assistance in editing the papers for publication. John Voss was a close collaborator in every phase of the project; the structure of the symposium and the contents of this volume bear the mark of his insistent and critical advice.

We also wish to express our warmest thanks to the officers and staff of the Johnson Foundation and particularly to Leslie Paffrath, Rita Goodman, and Kay Mauer. In recent years they have been the hosts at Wingspread for many Academy symposia; we have come to look upon them as close colleagues in the effort to clarify and illuminate issues at the interface of science and public policy.

The substance of the essays in this volume owes much to the discussions at Wingspread. In addition to those already named, the following individuals contributed memoranda, commentaries, and criticisms at the symposium: Stephen Berry, Qui Won Choi, Manfred Cziesla, Steven Feldman, A.A. Fouad, José Goldemberg, Christoph Hohenemser, Amory Lovins, Carlos Mallmann, Peter Meincke, D.L.O. Mendis, Roderick Nash, Y. Nayudamma, Kojiro Nishina, Jozsef Nyilas, David Pimentel, and Roger Revelle. In our three days of meetings, the symposium participants exchanged views and provided insights that advanced the debate surrounding appropriate technology beyond the bounds of national concern to a level where technological needs and human values could be discussed on a worldwide basis.

F.A.L.
A.O.

Appropriate Technology and Social Values— A Critical Appraisal

NA

*

Book title:

1-8

Introduction

Franklin A. Long

In the series of formal and informal discussions that led to the chapters of this volume, the initial focus was on the topic, technology choice and social values. The hope was to illuminate the way in which social values do and should influence technology choice by nations and by groups within nations. Most of the participants in the discussions believe strongly that social values should play a substantial role in the selection of technologies. Since both political objectives and social values differ considerably among nations, the implication is that these circumstances should dictate different technology choices—that is, the appropriateness of technologies should not be decided on purely economic and factor endowment grounds. This line of discussion led to a more direct consideration of just what is meant by appropriate technology, a term that has become something near to a battle cry in the debate surrounding technology and values.

The simplest meaning for the term appropriate technology, and the one used by most of the authors of this volume, is the tautological one which simply states that appropriate technology is the technology that is appropriate to the particular situation faced by a given group of people, with consideration given not only to economic circumstances and available resources but to value priorities. However, many thoughtful students adopt much more specific meanings, and the meanings have sometimes differed considerably between groups surveying technologies for the more wealthy developed nations and those considering technologies for the less developed. Thus to many analysts of the developed countries, appropriate technology is "soft" technology or technology "with a human face." For those concerned with the developing nations, the term is often used synonymously with "intermediate technology"—that is, technology that is midway between the traditional village technologies and the advanced capital-intensive technologies of the Western world.

The question of values is implicit in all of these definitions, whether applied in the developed or the developing world. The problem of proper response to values is of particular importance to the developing nations where the pace of development is frequently fast and where the value conflicts between the traditional rural societies and the Westernized urban elements are often very great. There is little disagreement on the basic social values to which people subscribe: security, social justice, equal opportunity, freedom for the individual, health and education for all, minimization of inequities in income. The problems for most developing nations lie in assigning priorities and committing scarce resources; hence the coupling of social values and technology choice becomes crucial. Development and use of new technologies means change, but change can either enhance the movement toward desired social values or, unhappily, negate it.

The authors in Part I of this volume deal with the ambiguities and uncertainties that surround appropriate technology and consider the ways in which values enter into technology choice. In his chapter, Paul De Forest reviews the various meanings of the term appropriate technology in the developed and developing nations and considers its role in "an increasingly polarized debate between advocates of 'low' versus 'high' technology." De Forest maintains that appropriate technology avoids this either/or choice by focusing on the mixed nature of nearly all technological systems whatever the level of development. For De Forest, the real significance of the term lies in the fact that it permits technologies to be evaluated in terms of criteria that range beyond cost, efficiency, and acceptability in the marketplace to include suitability to particular social contexts and consistency with desired social goals.

Langdon Winner carries this discussion further by presenting what is essentially an intellectual history of appropriate technology with special reference to the United States. His chapter traces the concept of appropriate technology back to nineteenth-century utopians such as Robert Owen and socialists like Thomas Carlyle and then examines in detail the flowering of the concept in the radical social movements of the late 1960s; in the writings of Lewis Mumford, Paul Goodman, Herbert Marcuse, Theodore Roszak, and Jacques Ellul; and in the emergence of such phenomena as the New Alchemy Institute. As Winner sees it, a prime consideration for most Western proponents of appropriate technology is a desire to find technologies more responsive to human needs than the current ones. Probing deeper, Winner concludes that the real issue for these critics is the stress and strain of modern Western civilization, which they attribute to the destructive nature of complex and impersonal technological systems. The chief puzzle to Winner is, why pick on technology: why not look for modified political and economic systems or even a new religion?

Harvey Brooks offers his own critique of appropriate technology. Brooks recognizes that the term has been used as an organizing concept by both the developed and developing nations, frequently with rather different meanings. His attention, however, is particularly drawn to usage in the developed Western world, where, with Winner, he concludes that the dominant meaning is that of alternative technologies more suitable than the existing ones, which are seen by critics as "hard," centralized, environmentally degrading, and socially impoverishing. Brooks is decidedly skeptical of this extreme vision of the present. He notes that a large, centralized, and hierarchical production system is often a prerequisite to a decentralized and non-hierarchical social system and mentions, as an example, the production and use of the automobile. Brooks is also skeptical of the economic attractiveness and ecological benefits claimed for some of the alternative technologies. He argues that were "soft" technologies truly superior economically, they would have already been chosen by a free market. Brooks' ultimate conclusion is that there is a case to be made for the symbiotic coexistence of some of these alternative technologies with some of the current Western technologies; he specifically suggests that there are technicoeconomic "niches," analogous to ecological niches, in which some of these alternative technologies can thrive and urges more vigor in searching for them.

John Montgomery focuses in his chapter on the link between technology and values and gives primary attention to the problems encountered by the poorer developing nations in making technology choices. He argues that values inescapably enter into these choices and that the technologies are not and cannot be value-neutral. The important questions as he sees them become, what values and whose values? Montgomery presents eight values that virtually encompass mankind's range of concerns, and he notes that each of these values has a positive and a negative side—that is, benefits to some and costs to others. Furthermore, social, political, and technological choices will almost routinely bring some of these values into conflict. Hence, the problem of "whose values" becomes crucial.

In the developing nations especially, the cleavage between the national elites and the poorer citizenry can be very great. Think, for example, of the deep differences between the rural villagers and the urban elites of India or Egypt. The operational problem is that all too often the value priorities of the ruling elites are the determining ones, and these may be very different from those of the poorer and politically subservient groups. In this situation, Montgomery concludes that if technologies are not socially neutral, neither are the scientists and technologists who develop and espouse technologies and who must inevitably accept some social responsibility for the technology choices that are ultimately made.

The essays in Part II of this volume turn more explicitly to the problems of technology choice in developing nations. The first chapter of the group, that by Gustav Ranis, firmly accepts a tautological definition for appropriate technology—namely, that appropriate "depends" on the time, on the place, and on the circumstances. Ranis then asks, what makes technologies appropriate, and how are developing countries to identify, select, and adapt potentially appropriate technologies to their particular needs? In most instances, the spectrum of technology choices available to developing nations is relatively wide; yet the technologies actually selected and used often appear to be particularly inappropriate. In analyzing the causes of this anomaly, Ranis points out that the character of the demand for technology in many of the less-developed countries (LDCs) does not easily lead to selection of the most appropriate, in large part because of the distortions induced by import substitution policies of the kind almost invariably promulgated within these nations. A particular consequence of these distortions is discrimination against agricultural output, which in turn tends to stifle demand for appropriate basic goods. A second difficulty is simply lack of information—first, about the alternative technologies or "quality bundles" of techniques that are available and, second, about what can be done to modify technologies to make them more appropriate.

On the supply side, Ranis notes that although the developed world has a wide selection of technologies from which to choose, this breadth is not often apparent when looked at from within a less-developed nation, partly because full information on technology alternatives is not usually at hand and partly because the capacity does not often exist for the detailed analyses needed to reveal the suitability of available technologies. Equally significant is the fact that the demand for technology stems principally from the entrepreneurs who propose to use it, yet it is difficult to insure that full information on technology alternatives reaches down to them. In other words, on both the demand and the supply sides, initiative from within the less-developed nation is essential in searching for, identifying, and ultimately selecting appropriate technologies.

Ranis underlines the important point that in generating the demand for technology, in developing adequate information on supply, and perhaps even more in appreciating what can and should be done to modify imported technologies, most less-developed nations have more to learn from each other than from the developed world. This argues for more regional collaboration among the LDCs. In his conclusion, Ranis expresses considerable optimism that reasonable policy actions can be taken by both the developed and developing countries to render technology choices by the less-developed nations more appropriate.

South Korea is one of the success stories in rapid growth and development by effective selection and use of technology. In the first of four case studies of technology choice in the developing world, H.S. Choi states that Korea deliberately made two crucial decisions: first, it has pursued economic growth by building up its industrial capability for export and, second, in developing these exports it has chosen the path of high technology—that is, using technologies obtained almost exclusively by importation. The implementation of this strategy has depended on an extensive and continuing interaction between a vigorous private enterprise sector that imports the technologies and manufactures the products and a strong governmental planning, monitoring, and guiding program that influences technology choice by means of a complex mixture of subsidies and regulations. As one element, Korea has built up an extensive and effective infrastructure of government institutions for information gathering, standards setting, technology analysis, and the education of skilled workers and professionals. Simultaneously, technology-licensing procedures have been established to help insure that technologies are obtained with appropriate characteristics and under appropriate terms. Concessionary government loans to industry have helped guide the development in desired directions. The consequences have been an extremely high rate of growth and a gradual shift from "bundled" technologies—for example, turnkey production units—to direct licensing of the central elements of the chosen technologies.

At least as impressive as the success of these industrial programs is the fact that South Korea has also managed, again through substantial governmental intervention, to advance the economic growth of the rural sector of the country at about the same rate as that of the urban so that no significant difference in family income exists between the two sectors.

That success in development is not automatic, even with extensive government intervention and guidance, is illustrated by the analysis of technological needs in Ghana by Robert Dodoo. According to Dodoo, the current Ghanaian five-year development plan specifies three basic goals: "capturing the commanding heights of the economy and placing them firmly in the hands of Ghanaians through the deployment of state power; promoting a stable economic growth in development; and insuring that the fruits of development are equitably distributed to improve the quality of life for all Ghanaians." In spite of these explicit goals and very considerable efforts on the part of agencies, many of the technology applications within Ghana remain backward and inefficient. This is strikingly true of agriculture, but it is evident in other areas as well. There has been no lack of good ideas and interesting programs. The failure appears to have been in the execution. Dodoo gives no final explanation for the failure but points to several areas of difficulty. Problems are posed by the inadequate linkages between the various policy and action groups within the country; in particular, the scientific and technological activities are relatively isolated and the R&D laboratories that have been established have been ineffective in promoting the utilization of their output. Shortages exist in skilled manpower. More emphasis is needed on modified traditional technologies that would be appropriate for the rural sector. Perhaps the principal message from this analysis of Ghana's difficulties is that lack of progress in any one of the several components of development can seriously constrain the total effort.

A particularly interesting analysis of differing strategies for development is given by V.V. Bhatt who discusses two alternative approaches to development in India—the Sarvodaya (Gandhi) and the alternative socialist (Nehru) strategy. For Gandhi, the overriding problem in India was poverty, which had to be tackled directly by providing more employment in traditional sectors—that is, in agriculture and rural industry. In sharp contrast, the Nehru strategy gave priority to large-scale, government-owned industrialization and rapid development of a modern sector based on science and technology. Bhatt attempts an evenhanded analysis of the two strategies, listing the virtues and the difficulties of each. However Bhatt is persuaded that lack of attention to the rural sector, which need not have been but was an element of the Nehru strategy, has been a mistake, and he is clearly pleased that the current Indian government is giving stronger emphasis to development in the rural areas. Bhatt's strong conviction from this comparative analysis is that development strategies do make a difference. He is convinced that for an agricultural country like India, assigning a high priority to rural development and full employment is the central component of a successful strategy.

In the final case study, Dwight Perkins analyzes China's experience with rural small-scale industry. Perkins cites a number of reasons why it was almost predetermined that China emphasize rural development—the critical need to maintain adequate food supplies;

the virtual lack of transportation to the rural areas, which required the decentralization of much supporting industry; and the lack of capital and skills for development of large-scale centralized technologies. In Perkins' view, China's emphasis on small-scale rural industry has been successful, not only in the minimal sense of staving off starvation and chaos, but also in being in many cases more cost-effective than large centralized production units. At the same time, it has led to a number of other important benefits, among them the widespread dissemination of knowledge of modern (if not necessarily highly advanced) technologies, the changing and greatly expanding role for women in rural society, and the development of administrative capabilities at the commune and county levels. By maintaining employment in the rural areas, China has been able to avoid some of the worst aspects of the urban-rural polarization that characterizes so many developing nations. Although many of the specific elements of China's programs are unique to it, Perkins believes that the basic strategy of rural industrialization is applicable to many nations and would be of major importance to most of them in reducing the discrepancies between rural and urban areas.

All of these discussions of individual LDC experiences demonstrate that national development is a complex, multifaceted process that calls for simultaneous action on a variety of fronts: improved education and public health, financial planning, provision of a supporting physical infrastructure, industrial development, and maintenance of an appropriate balance between agriculture and industry as well as between rural and urban development. To bring these elements along together calls for a continued coordination and information exchange among a wide variety of public and private sector activities. Another almost mandatory requirement is a continuing national planning effort that is consistent with national goals and objectives.

Implicit in this listing of the elements of the development process is the considerable role that social values should and usually do play in development strategies generally and in the selection of appropriate technologies in particular. Full employment, improvement of the lot of women, and a minimization of the inequities between rural and urban regions are all value-laden goals, and the priorities given to them will clearly influence the appropriateness of a given technology. But the operational problem of successfully inserting consideration of these and other values into the process of technology choice remains a thorny one.

In the wide-ranging final chapter of this volume, Kenneth Boulding approaches the question of being rich and being poor from an almost philosophical point of view. He notes that one of the great puzzles in human history is why increases in productivity have taken place in some societies and not in others. It was once freely assumed that national development would be relatively easier for the newly established, poor nations of Africa and Asia to achieve, in that they not only had access to the many technologies of the wealthy nations, but also had the example of how these nations had accomplished their development. Most of us now find this argument a very dubious one, even to the point of wondering whether the existence of such striking examples of development as the United States and Japan is not an inhibiting factor rather than a helpful one. In Boulding's words:

> The rich and the poor represent different ecosystems of human knowledge and artifacts. Development into riches is not just "growth," it is a profound change in the knowledge structures and behavior and skills of the population. . . . The eradication of poverty and the development of a world that is reasonably uniformly rich is an objective for the human race that commands our very deepest assent. But the fulfillment of it is difficult and may take a painfully long time.

If this volume can help even slightly in easing the path and shortening the time to this central objective, the contributions of all those involved in its preparation will have been worthwhile.

Part I

Appropriate Technology as Concept

Chapter One

Technology Choice in the Context of Social Values— A Problem of Definition

Paul H. De Forest

11-25

The term *appropriate technology* is now commonly used in discussing the relation of technology to social values.

Unfortunately, the analytical potential of the concept has not yet been realized, in large part because the term has been captured by one side in an increasingly polarized debate between advocates of "low" versus "high" technology. This chapter explores the reasons for the occurrence of this development, assesses its implications for the relation between technology and social values, and advances some suggestions whereby the concept of appropriate technology might more constructively be employed in analyzing technology choice in both developed and developing nations.

TOWARD A DEFINITION OF APPROPRIATE TECHNOLOGY

The terms used to define the relation of technology to society can have precise meanings (see Appendix, p. 23), despite the tendency toward ambiguity and imprecision in their normal use. Within this conceptual structure, the term appropriate technology has significant analytical potential in that it can contribute to an understanding of the complexity of the content of technology and can also promote the incorporation of social values into the process of technology choice.

Appropriate technology has no intrinsic meaning, but can only be understood in relation to specific social, economic, and cultural referents. Other commonly used concepts connote more narrowly limited dimensions of the content of technology. For example, the continuum from "sophisticated" to "simple" technology describes the research and development base, that from "advanced" to "traditional" the quality of engineering, that from "large-scale" to "small-scale" the level of industrial operation, that from "centralized" to "decentralized" the focus of societal organization, and that from "high-cost" to "low-cost" the quantity of capital inputs. Precise definitions of the endpoints of each continuum and the distance between them vary greatly from one society to another, depending on the state of development. The choice of a given continuum tends to channel analysis of technological development in a single direction, often to the exclusion of other important dimensions.

Some critics have placed the blame for the ills of modern industrial society on "advanced technology," as if it subsumed the attributes "large-scale," "centralized" and "resource-intensive." Others have urged reliance on "intermediate technology" to solve the economic and social problems of the developing nations, arguing that "advanced technologies" inevitably increase centralization by driving people from their villages, raise unemployment by replacing labor with capital, and destroy the quality of the social environment and traditional systems of values. Yet there exist advanced technologies, measured in terms of the quality of their engineering, that are also small-scale, decentralized, and resource-efficient. A solar-powered water pump is one such technology that is useful in both developed and developing countries. The appropriateness of the solar-powered pump can be measured in terms of several criteria considered together: scale, cost, localized control and efficiency, and engineering state-of-the-art.

When one emphasizes its potential contributions to the quality of life, a different dimension of appropriate technology comes into focus: the incorporation of social values into the process of technology choice. Technologies may be judged appropriate not just in terms of their level of sophistication and complexity, but also with regard to their suitability to particular social contexts and their consistency with desired social goals. The judgment of appropriateness requires a choice among values, and since values are often in conflict, the selection of appropriate technologies is deeply embedded in the political process. Implicit in the use of the concept "appropriate technology" is the presumption that the form, scale, and function of technology are flexible and malleable—able to be shaped, defined, and directed by organized social agencies.

In sum, appropriate technology as *process* has generally been defined as a fundamental alteration in the procedures whereby technologies are selected and implemented in order to give greater weight to social values such as decentralization and individual control and less weight to the relatively unimpeded operation of market forces. Appropriate technology as *content* has come to be applied to a special class of technology systems: those incorporating energy-efficient, labor-intensive, small-scale, decentralized techniques. Just as "technology" has become for many synonymous with the body of the most advanced and sophisticated techniques, as if "modern technology" were redundant and "traditional technology" contradictory, so "appropriate technology" has tended to be coopted by the critics of modern society to be used interchangeably with "alternate technology" or "intermediate technology," as if "advanced technology," or the existing pattern of technology-in-use in a given setting, were somehow by definition inappropriate. Each case is a misuse of a potentially richer and much more flexible concept.

APPROPRIATE TECHNOLOGY FOR
THE DEVELOPED NATIONS

The extent to which the concept appropriate technology has been incorporated into the critique of modern industrial society is epitomized by the perspective set forth in Langdon Winner's *Autonomous Technology*. Winner's conception of the relation of appropriate technology to social and political values is encapsulated in what he regards as a "simple yet long overlooked principle: *Different ideas of social and political life entail different technologies for their realization.*"[1] Yet social and political values do not, he believes, inevitably produce technologies appropriate to them; on the contrary, technologies can become so deeply entrenched that they can erode traditional or existing values and shape new ones. Appropriate technology, then, is technology put in its place, subordinated to the central values of individual or social freedom and the right to choose. When Winner defines this approach as epistemological Luddism, he is not so much begging off from a physical confrontation with the machine as delineating his primary target: the social, economic, and political value systems of modern industrial society which have interacted with technology systems to increase the feeling of powerlessness, promote the domination of elites and contribute to the aggrandizement of the technological myth. Winner is reluctant to propose specific technology systems appropriate to his vision of an alternate society. A number of approaches toward restructuring technology systems in modern industrial societies have been proposed by others, however.

The most extreme critics of modern technology, including Charles Reich[2] and Theodore Roszak,[3] argue that nothing less will suffice than the dismantling of advanced technology and its restructuring along fundamentally different social and cultural lines. Whatever one's opinion of the merits of the counterculture movement, its target is clearly not technology per se but the political and economic order that shapes and directs technology according to the interests of a ruling elite. Barry Commoner's evaluation of energy technologies in his *The Poverty of Power*[4] is consistent with this philosophy. It will never be possible, he argues, to resolve the energy crisis without first thoroughly reordering an economic system that, through its pricing mechanism, promotes inefficient use of energy and degrades the environment. Without a redefinition of the concepts of energy and economy, such measures as conservation, taxing policies, and the use of more efficient energy technologies will have little chance of success. Given such assumptions, determining which technologies are "appropriate" will be dependent on more fundamental decisions regarding appropriate economy and appropriate polity.

Another group of critics, greatly influenced by the Marxian notion of the alienation of man from technology, focuses its attention upon particular aspects of the role of technology in modern industrial society that are held to be inappropriate to the rightful position of man as master of technology. Following Marx, these critics hold that modern industrial society has directed technology toward the liberation of production from the constraints of human strength and stamina rather than the liberation of mankind from the limitations of nature, so that man has become increasingly subordinate to the machine. Among the manifestations of modern technology believed inappropriate are centralization of technological organization which limits participation and control by affected individuals and local communities; increasing scale and complexity of particular techniques, tools, and machines which make worker participation in management decisions more difficult; and growing elaboration and sophistication of technology systems which serve to reduce the comprehensibility of products to consumers.

Rejecting the argument that advanced technology inevitably produces centralized control, Murray Bookchin[5] asserts that his notion of "liberatory technology," which requires a decentralized economy and institutions structured on a human scale, enables economic and political control over technological form and function to be held by local communities. In his plan, simple, small-scale technologies will meet most local needs. Advanced automated technologies that are readily portable or able to be operated locally (computer terminals, for example) will be widely used. Larger-scale technologies will be relied upon where necessary or advantageous, but they will be supplemental rather than primary. Liberatory technology can be viewed as an example of appropriate technology as defined here since it takes into account a range of qualitative factors, it stresses the necessary relation of technology to social values, and it is deeply embedded in the political process. The problem with this, as with other broad schemes, is not so much the appropriateness of technological structure to value system as the appropriateness of value system to social setting. It might be doubted that so mixed a bag of technologies could ever be structured into a system, but the more basic question is whether the radically restructured society required for a liberatory technology is foreseeable given the political and cultural patterns of most advanced nations. The "crisis of the machine," which is so real to the critics of modern technological society, has received little attention from policymakers and the public. Those who argue on behalf of the appropriateness of decentralized technologies—the capacity they provide for local control, their liberatory potential, their ability to promote the unity of man with nature—are often regarded not as the vanguard of social and economic reconstruction but as echoes of an already outdated counterculture movement or as throwbacks to an even earlier tradition of small, self-sufficient, utopian societies totally out of place in the industrial age.

Another group of critics whose perspectives on appropriate technology deserve a hearing arise, not from an antitechnological ideological tradition, but from the technoscientific community itself. Their criticism is rooted in what they see as a trend toward ever larger and more complex and interdependent high-technology systems, systems that not only increase the dependence of society upon technology and subject the individual to the vagaries of centralized bureaucracies but are also becoming increasingly less efficient and outdated when measured against the frontiers of scientific and engineering disciplines. There exist, they assert, more appropriate systems of technology at least as advanced and sophisticated as the present systems but even more efficient in their use of material and energy, less subject to strain and breakdown, and more compatible with such social values as individual freedom and control. Their proposals are not directed toward the alteration of fundamental value systems or the reshaping of technological structures, but instead toward specific functional areas where the pattern of technology-in-use is held to serve inadequately such social needs as energy, pollution control, food production, and health services.

In perhaps the most publicized set of arguments for different and more responsive technologies, Amory Lovins[6] has proposed a group of "soft" energy technologies as more appropriate to ensuring social and economic stability and growth into the twenty-first century. A wealth of argument has been developed to support the viability of these decentralized, energy-efficient and socially valuable technologies. Yet the proposals have generated considerable opposition. On the one hand, Lovins has been accused of being selective in his data, too optimistic regarding the efficiency and savings that would be generated, prejudiced in favor of the appropriateness of soft energy technologies to the social goals he postulates. On the other hand, it is asserted that Lovins paints too grim a picture of the future prospects for hard energy technologies, and that the social, economic, and political reforms required to implement soft energy technologies are far more radical and extreme than necessary. Although Lovins may be right in maintaining that soft energy technologies represent an important direction for long-term development, he probably underestimates the social and political difficulties that stand in the way of a larger near-term role.

Calls for sweeping reforms in the existing pattern of technology in a given sector of advanced industrial societies tend to minimize the complexity of the social and economic interdependence of the total society into which systems of technology are deeply integrated. The implementation of soft energy technologies on a wide scale would pose immense problems of adjustment in other areas of society, affecting industrial production, the housing industry, transportation, styles of living and working, and rates of inflation and employment. It is not enough to claim the theoretical feasibility of appropriate technologies for low energy and continued economic growth. There must also exist a realistic strategy for adoption of such technologies and the resolution of the strains they will cause in society at large.

Disturbed by the apparent political and economic advantage of groups wedded to the status quo and impatient with the resulting obstructions and delays, many advocates of appropriate technology look for its adoption through radical, sweeping changes produced primarily in the political arena. One should not be too sanguine, however, about the effectiveness of the polity as the primary focus for the adoption and implementation of technological innovations. The weight of historical and contemporary evidence is heavily in favor of the marketplace as the arena where the most viable decisions regarding technological form and scale are reached, whatever the nature of the economic or political system. From a moderate, less apocalyptic perspective, the record of entrepreneurs, managers, and consumers is not that bad—smaller, more efficient, cleaner automobile engines; better insulation and more economical heating systems for houses and commercial buildings—the list could be extended indefinitely. The political system can provide incentives—and disincentives—that promote appropriate decisions by individuals, but its ability to legislate such decisions is limited.

What conclusions, then, can be reached concerning the future of appropriate technology in the advanced nations? First of all, the ability to gain the information and understanding upon which decisions regarding appropriateness must be based is improving rapidly. The data consist of both technological indicators, generated by the tools of technology forecasting and assessment, and social indicators, that measure the state of society, its quality of life, and its rate of progress toward social goals. What remains problematic and subject to dispute is the character of the relation of technological indicators to social indicators. This relation is central to appropriate technology; it is the reason that definitions of the term differ widely and judgments concerning the appropriateness of specific forms and patterns of technology remain controversial. Consequently, acceptance of appropriate technology as an analytical and evaluative criterion would amount to a recognition of the fact that the form and scale of technology in society is open to question, that there is more than one potential system of technology for the advanced industrial nations.

In the ongoing debate in the United States and other Western nations, however, the term appropriate technology has been coopted by the critics of modern industrial society for whom the questions have already been answered. Although there are as many kinds of appropriate technology as there are arguments in the critique of modern society, most schemes are nonetheless sweeping in their proposals for alternate technologies and in their calls for social change. They are, in essence, ideological and political—the attempt to enact one person's or one group's social vision. Here, appropriate technology becomes a slogan for the implementation of radical social goals, truly a weapon in a political struggle.

APPROPRIATE TECHNOLOGY FOR THE DEVELOPING NATIONS

The potential for controversy over the meaning of appropriate technology and its relation to social values is at least as great in the developing nations as in the developed nations. The definition of social and economic goals, their impact upon traditional value systems, and the form and scale of technology appropriate to them are matters of serious debate. The distinction between developed and developing nations relates to the process whereby appropriate technologies are selected and implemented. In the developed nations, decisions regarding technological form and function, whether the product of central planning or free market choices, are primarily domestic. By contrast, most developing nations are heavily dependent on outside technologies and are hence influenced by the political and economic goals of technological aid programs shaped abroad. Traditional programs of development assistance, concentrating on high-cost, large-scale, politically visible projects requiring centralized administration, have recently been subjected to severe criticism. A new approach to foreign aid, captured in the notion of "basic human needs," is becoming increasingly popular. If past large-scale technological aid programs have been ineffective, the argument runs, perhaps it is time to strike out in a new direction, toward greater emphasis on technologies appropriate to basic economic and social needs at local levels, where the people are concentrated and the deprivation is greatest. Officials in the developing nations are, however, reluctant to accept a definition of appropriate technology that focuses only on the small-scale needs of rural communities. The solution to the shortcomings of technology aid programs can be found, they argue, not in the alteration of the content of technology but in reformed procedures for technology transfer, including greater use of multinational channels and more equitable policies in the exporting nations, to ensure that the needs of the recipient less-developed nations are given primacy. Some have argued that "self-reliant technology," defined as control by the developing nations of the decisions on what technologies are developed or imported and how they are used, is the most appropriate system of technology to promote independent development.

What are the chances that the primacy of domestic values within systems of self-reliant technology would produce technology systems more appropriate to the goals of the developing nations than those currently in existence? With certain exceptions, for example, China and Tanzania, the content of the technology systems would probably not be radically transformed. The international economic system is becoming increasingly integrated, and the pressures toward convergence intensify as a nation modernizes. The tendency to select high-technology systems, capital-intensive investment, even imported machinery and products, is pronounced throughout the developing world. Preferences for high-technology systems can be seen to be desirable and logical; yet they all are based on Western notions of productivity and development and share in the dominant Western values.

However, from the perspective of the poorest and most remote regions and communities, where traditional values are seriously challenged by nationalization and established patterns of production and consumption are being eroded, the appropriateness of such choices may be seriously questioned. For many of these regions, centralization and modern technology have done little to arrest the slide toward idleness, hunger, and social dislocation. What technologies might be more appropriate?

Considerations such as the foregoing have been incorporated into the concept of intermediate technology, popularized by the British economist E.F. Schumacher.[7] What is needed to improve the lot of the poor, Schumacher argues, is an emphasis on the Gandhian system of village technologies whereby workplaces are located in the areas where people live, are dependent on the locally available labor supply and are directed toward the satisfaction of basic human needs. Clearly, in this view, lifestyle and livelihood are intimately related: work, especially creative work, is a psychological, even a physical necessity. Technologies appropriate to the availability of work and the psychological rewards that flow from creative work are small-scale, labor-intensive, decentralized, intermediate technologies.

Despite the arguments of advocates of intermediate technology, there is nothing so inherently valuable about traditional technology systems as to justify their being frozen in time and space. In almost all cultures and nations technology has historically been reflective of rather than consciously designed to promote social patterns and values. The content of technology systems is dynamic in nature, changing and evolving with new experiences and expanded opportunities but also dramatically affecting social values, broadening cultural horizons, and sometimes displacing long-established traditions. The poorest people are not immune from these forces. They seek more of the "goods" of life and more leisure to enjoy them. Symbolic slogans such as "small is beautiful" and "Buddhist economics" can contribute little to the ability to satisfy this revolution of rising expectations. Nevertheless, contrasts in traditions and value systems, as well as in economic and technoscientific capacity, are far more extreme among regions and between urban and rural areas in the developing world than in the industrialized West. This situation argues for the very diversity and freedom of local experimentation discouraged by both large-scale technology assistance programs from abroad and national plans for centralized technological development.

One can readily understand, therefore, why most indigenous efforts at creating appropriate technologies have been conducted by small, local, experimental groups, whether public or private, whether motivated by economic profit or social benefit. As Nicolas Jéquier documented in his study for the Organization of Economic Cooperation and Development,[8] a large proportion of such efforts have been initiated by small private entrepreneurs whose motivations were eminently "practical": increased employment, utilization of local capital and materials, improved production and consumption and, ultimately, profit rather than loss. To be appropriate in a given setting and for a specified purpose, technologies will first have to be "salable" to their intended markets—providing sufficient jobs, operating efficiently in the local context, producing products desirable and affordable to consumers and profitable to their investors. Technologies not directed toward satisfying needs and desires beyond a minimum subsistence level will not be able to meet these criteria: leisure activities must be guaranteed as well as jobs; a few of the luxuries of life must be provided as well as the necessities.

Some respected analysts doubt whether small-scale, labor-intensive, intermediate technologies can meet these requirements on a wide scale and in the long run. In an exhaustive survey for the U.S. National Academy of Sciences, Richard Eckaus[9] found little evidence that village-based technologies could provide constructive employment and attractive products, much less promote the essential goal of national economic development. As Eckaus points out, the criteria by which appropriateness is usually measured are internally inconsistent; for example, maximizing consumption depends on capital investment, which in turn reduces employment. The one dimension that he found to be most appropriate was substitutability between capital and labor, determined by factor costs and profit–loss prospects.

Finally, if there is to be any real hope for eventual modernization and development, major emphasis will have to be devoted to sophisticated technologies for energy generation, industrial production, and more efficient agriculture. Solar pumps can improve water supply and solar cookers can save scarce reserves of wood and coal. But for the near future they will not replace dams and electrical power stations. Also, given the immense number of villages and consumers, heavy industry and large-scale technology will be required to produce sufficient numbers of solar pumps and cookers at an affordable price. What is required is a sense of the appropriateness of a given technology within a large and complex economy and society.

J.E. Goldthorpe stresses the need to keep small-scale technologies in their limited, if significant place:

> While the arguments for intermediate technology may appear very cogent, they seem to offer no escape from the dilemma that whereas "capital-intensive development" may mean high wages for the few and unemployment for the many, "labour-intensive development" is a polite way of

saying low wages. For example, in a recent television program on this problem a hand press was shown operated by about a dozen men in a small cooperative workshop producing hubcaps for automobiles. Though this was praised as an example of self-help and intermediate technology, it may be doubted how much was added to human welfare and dignity by a dozen men in a South Asian country working repetitively for several hours a day to produce what a machine in Detroit or Coventry could turn out in a few seconds.[10]

One might agree that a hand press is an inefficient, not very digni-fied, and certainly low-paying method to produce hubcaps. Yet if one examines the overall system of production from the perspective of local as well as national needs, the picture might become more acceptable, even attractive. The hand presses will themselves have to be manufactured, presumably at a more central location in a larger facility. The hubcaps will be placed on automobiles, to be assembled centrally but distributed locally—hopefully to some of the press op-erators. The automobiles will be driven on roads constructed and maintained by locally employed labor. Development is a long and complex process in which a range of technologies (high to low, hard to soft) will in various settings and at different stages be found to be appropriate.

APPROPRIATE TECHNOLOGY IN
AN INTERNATIONAL CONTEXT

In the worldwide debate over the relation of technology choice to social values there exists a real danger of misperception and mis-judgment, the result of the frequent use of two extreme and con-tradictory frames of reference. For many of those who adopt the perspective of the prosperous nations, the dominant concepts have become those of balance, moderation, stability, and marginal change. With reference to the industrialized Western societies, those individu-als advocate preserving the basic pattern of large-scale technology but with increasing attention to improved efficiency, planning and control, and the maintenance of individual choice. At the same time, they maintain that the developing nations should scale down their expectations for rapid economic growth, modifying Western tech-nologies to make them appropriate to changing social needs and val-ues. In contrast, for increasing numbers of spokesmen from the poorest nations, the dominant concepts are related to much more sweeping alterations in the status quo. The industrialized societies have a responsibility to halt or even reverse their pace of growth and to promote global equity by providing much more extensive economic and technological aid to the developing nations. The latter will then be able to mount crash programs toward social and economic modernization, based in large part on rapid technological change under their own self-reliant control and direction.

Is there any way in which utilization of the concept of appropriate technology can help to alleviate this implicit conflict? Advanced technology has permeated the social and economic fabric of the industrialized nations, measured by such indicators as education, GNP, resource consumption, and research and development investment. The developing nations, by definition, rank far lower on these indicators of economic and technological advantage. The spread of intermediate technologies at the village level might have a dramatic impact on the quality of life while making little difference in the indicators of growth and development. How would this affect the perception of relative disadvantage shared by decisionmakers and public alike? The average inhabitant of advanced nations will still consume many times his share of global resources and will still rank far higher on most indicators of prosperity. How long could the persistence of this gap be tolerated? From a different perspective, what specific techniques or technological systems might, if implemented in advanced and developing nations alike, be tolerable to both—that is, economically acceptable, culturally consistent, and supportable by the public at large—while at the same time promoting the cause of global equity?

If one were able to depart somehow from an either-or-choice among "high" and "low" technologies and to affirm the truly flexible nature of the concept of appropriate technology, a way out of the dilemma might emerge. The real significance of appropriate technology is as an indicator of the "mixed" nature of most technology systems: that they include, as an inevitable result of the complex role they play in society, some components that are "high," others that are "low," and still others in the middle of the scale, measured in terms of such criteria as size, centralization, capital investment, resource inputs, and engineering base. In the advanced nations almost all technologies are both centralized and decentralized depending upon their position in the flow from innovation to adoption to adaptation to implementation, or from resource to production to distribution to consumption. Likewise, in the developing nations technologies suitable to employment, economic growth, and development will necessarily include a mix of levels of technological sophistication, of labor and capital requirements, and of range of skills in design, manufacture, and dissemination.

What, then, is the potential for global convergence through the use of appropriate technologies? In brief, it is only as great as the potential, on the one hand, for convergence in social, economic, and political systems, or, on the other hand, for transnational consensus both on the problems facing mankind and on the nature of technologies that, if implemented globally, would contribute to their solution. Given the severely limited prospects for true convergence in economic forms, social structures, political systems, and scientific and technological capacity, the greatest likelihood for the global implementation of common appropriate technologies is in specifically delimited areas—energy conservation and efficiency, environmental preservation, population control, and, perhaps most important, international security and arms limitations—where the widespread acceptance of such technologies is universally perceived as furthering global concerns.

APPENDIX

Discussion of the relation between technology choice and social values has been clouded by the use of terms that are defined loosely, if at all, and tend to serve more as symbols than as substantive concepts. The problem begins with the basic term "technology" and extends to the attributes of technology systems and contrasts among technology systems. The definitions provided below are used consistently throughout the text of this essay, but their use in other contexts may vary significantly from these definitions.

A. General Terms

1. Technology. A society's pool of knowledge regarding industrial arts and sciences, including the principles of physical and social phenomena and the application of these principles to production and the day-to-day operation of production.

2. Technique. A discrete item from the pool of technology in a society, e.g., a tool, machine, skill, process, etc.; may be either "hardware" or "software."

3. Technology system. A set of techniques organized and integrated according to criteria of appropriateness, including economic efficiency, cultural values, established policy, and the state-of-the-art.

B. Dimensions of Technology Systems

Components	*Parameters*
1. Focus of organization	centralized . . . decentralized
2. Scale of operation	large . . . small
3. Factoral inputs	capital-intensive . . . labor-intensive
4. Utilization of resources	
a. Intensity	heavy . . . light
b. Efficiency	efficient . . . inefficient
5. Level of development	modern . . . traditional
6. Research and development base	sophisticated . . . simple
7. Engineering state-of-the-art	advanced . . . intermediate . . . primitive

C. Categories of Technology Systems

1. High-technology systems. Tending toward the extreme on most or all of the following components: centralized, large, sophisticated, capital-intensive, modern.

2. Low-technology systems. Tending toward the extreme on most of the following components: decentralized, small, simple, labor-intensive, traditional.

3. Mixed-technology systems. Components of both high- and low-technology systems but with little systematic planning or coordination.

4. Intermediate-technology systems. Systems consciously designed to fall approximately midway between the extremes on most of the listed dimensions.

5. Dual-technology systems. Simultaneous operations of consciously designed high- and low-technology systems in different economic sectors or regions within the same society.

6. Alternative-technology systems. Sets of techniques organized and integrated according to criteria that are in fundamental contrast to those currently operating in a society on most of the listed dimensions.

NOTES TO CHAPTER ONE

1. Langdon Winner, *Autonomous Technology* (Cambridge, Mass.: MIT Press, 1977), p. 325 (italics in original).

2. Charles Reich, *The Greening of America* (New York: Random House, 1970).

3. Theodore Roszak, *The Making of a Counter Culture* (Garden City, N.Y.: Doubleday, 1969).

4. Barry Commoner, *The Poverty of Power* (New York: Knopf, 1976).

5. Murray Bookchin, "Towards a Liberatory Technology," in *Post-Scarcity Anarchism* (San Francisco: Ramparts Press, 1971), pp. 83-139.

6. Amory Lovins, *Soft Energy Paths*, (Cambridge, Mass.: Ballinger, 1977).

7. E.F. Schumacher, *Small Is Beautiful* (New York: Harper & Row, 1973).

8. Nicolas Jéquier, *Appropriate Technology: Problems and Promises* (Paris: Organization of Economic Cooperation and Development, 1976).

9. Richard Eckaus, *Appropriate Technologies for Developing Countries* (Washington, D.C.: National Research Council, 1978).

10. J.E. Goldthorpe, *The Sociology of the Third World* (Cambridge: Cambridge University Press, 1975), p. 297.

Chapter Two

Building the Better Mousetrap: Appropriate Technology as a Social Movement

Langdon Winner

As if to provide a landmark for the social movement called appropriate technology, the Nieman-Marcus 1977 Christmas Catalogue announced the sale of his and hers urban windmills. Available at a cost of $16,000 each, the devices would, according to the department store, enable customers to "enjoy today's electrical appliances and gadgets without overtaxing public power supplies, family utility bills, or tempers." The ad assured prospective buyers that the windmills were nonpolluting, noiseless, and environmentally safe. In a city that provides an average wind speed of twelve miles per hour, the mill "could generate more than enough wattage to brew her morning coffee, Benedict an egg, heat her hair rollers, soothe her psyche with stereo, and give her bronze beauty while she relaxes under the sun lamp." Other uses cited for the mill's energy include blending daiquiris and supplying power for junior's rock band.

Saturated by the influence of commercialization, advertising, and bureaucratic cooptation, movements for social change in the late twentieth century often become indistinguishable from fashion trends. In a matter of weeks the radical thrust of a new idea or practice can be absorbed into the ephemera of glossy surfaces of the postindustrial marketplace. It is possible, perhaps even likely, that appropriate technology will encounter this fate. "Wisdom demands a new orientation of science and technology towards the organic, the gentle, the non-violent, the elegant and beautiful,"[1] E.F. Schumacher argued in his wildly popular book. But who among us is equipped to overcome the institutional inertia of modern industrial society to realize this vision? True, the proponents and practitioners of appropriate technology—the New Alchemists, Faralones Institute, Intermediate Technology Development Group, engineers of the "soft energy path," and the like—show a genuine seriousness in their projects.[2] But the most likely result of their efforts can already be anticipated. As hopes for a new society founded upon a humane, relatively small-scale, decentralized, environmentally prudent technology wither into a set of lifeless slogans, the tangible fruits of appropriate technology—energy from various renewable resources, for example—will be picked up by commercial interests as salable items and adopted by public and private bureaucracies as flashy new "programs." Another possible source of opposition and change will have been incorporated as decoration in a society whose basic form and direction remain unchanged.

During the past decade the notion of appropriate technology was first proposed as a way of addressing the economic, technical, and social problems that face Third World countries. Here the idea took shape in a situation in which technological choices are, at least in principle, still wide open. If one's intention is to find a path to economic improvement in harmony with the conditions of traditional society—its level of skills, its available resources, its most pressing social needs—intermediate or appropriate technology appears sensible. Even the most enthusiastic modernizers can find some virtue in plans to develop small-scale industrial and agricultural operations like Schumacher's one-person egg-carton factory.[3]

Since the early 1970s, however, the idea of an appropriate technology has been applied to advanced industrial societies as well. Although the circumstances of less-developed and developed nations are clearly linked, serious conceptual and practical difficulties arise when one talks about appropriate technology in situations in which a modern economy and complex sociotechnical structures already exist. Questions as to what is or is not "appropriate" in such situations border on the perverse. For if one wants to argue that our present technologies of industrial production, agriculture, communications, transportation, and the like are somehow inappropriate, one is faced with the question of what to do with them. Should certain practices or institutions be modified? Dismantled? Replaced?

In the literature of appropriate technology, such questions are usually avoided. Its advocates speak of "a revolution in technology to give us inventions and machines that reverse the destructive trends now threatening us all."[4] But how this "revolution" would affect the existing technological stock is seldom mentioned. Thus, many appropriate technologists who want to apply their work to the environmental and social problems of the developed (or overdeveloped) world sometimes represent their activities as if they were intended for developing nations only. Even the now standard choice of the label "appropriate" for the kind of technology sought reveals this caution. A more honest appellation now generally avoided would be "alternative technology," which has the virtue of pointing toward a mode of technics tailored to a different set of principles about nature, humanity, and society than the modern age has thus far followed. But, evidently radicals have now agreed with such institutions as the Rockefeller Foundation and the World Bank that an "appropriate technology" is what is needed.

In the following pages I want to examine the near and distant origins of appropriate technology, trace both its obvious and less obvious motives, and identify its distinctive model of social change. My intent is neither to praise the subject nor to bury it. Instead, I want to identify the contribution that the appropriate technologists make, despite their many follies and frustrations, to the articulation of issues in public life. Whatever practical results their explorations in hardware, technique, and economics may eventually yield, they already tell us a great deal about a crisis of authority in our time—a crisis in which formerly sacrosanct institutions of modern technology have become the object of direct philosophical and political criticism.

POLITICAL AND INTELLECTUAL ORIGINS

As one seeks to identify the sources of concern about appropriate technology in advanced industrial societies, it is not possible to try to locate a tradition in the strict sense that one can find the origins of contemporary labor movements in working-class struggles of the past. The diffuse, disorganized character of activities of this kind requires that we concentrate upon certain resonances among ideas and social experiments that have no direct historical ties. Thus, depending upon whom and what one counts, sources of the idea can be found in the following: Robert Owen and other nineteenth-century utopians, William Morris, Peter Kropotkin, Gandhi, the Spanish anarchists, the guild socialists, the cooperative movement, Lewis Mumford and the American decentralists, the followers of Rudolf Steiner's biodynamic agriculture, back-to-the-land movements, and whole generations of tinkers and crackpot inventors.[5]

The more immediate sources of interest in a radical approach to technology, however, derive from social movements of the 1960s and early 1970s. Many of the same concerns and passions that fueled activism in civil rights, the New Left, antiwar protests, counterculture, and environmentalism led eventually to a critical reexamination of the foundations of modern industrial society. This was territory that few persons involved in the early stages of 1960s American radicalism were eager to enter. Civil rights and Vietnam War protests of 1962 and 1967 seldom identified either capitalism or technology as the cause of troubles. While the workings of the "machine" or "system" were frequently mentioned as deeper maladies underlying social injustice and international aggression, such terms were employed metaphorically and did not point to anything specifically technological. A readiness to focus upon machines, techniques, chemicals, experts, and large-scale sociotechnical systems as sources of evil arose later on as social conflict in the United States and other industrialized nations of the late 1960s intensified. Searching for a common element in the maladies that angered and vexed them, activists focused upon modern technology as an increasingly obvious and perhaps too convenient key. Campus demonstrations against American chemical corporations vented students' rage against the napalm dropped on helpless mothers and children in Vietnam. The lethal hardware of the modern battlefield, presented with shocking clarity each night on network television newscasts, called attention to the war's technical side. Those who could not bring themselves to study the roots of Vietnam in the workings of American foreign policy or the goals of international capitalism were able to focus upon the instruments of destruction and upon the persons who manned and directed the B-52s and helicopter gunboats. Soon the link between destructive herbicides in Southeast Asia and the herbicides, pesticides, and industrial pollutants so recklessly dispensed at home gave new meaning to the word "environment." For fully one and one-half centuries previous to the Vietnam War, the advance of technology and the horizons of human progress had seemed one and the same. Now it appeared that the most sophisticated products and best practitioners of scientific technology had somehow been enlisted in a war against progress and toward the annihilation of everything decent, humane, and just.

This obsession with technology arose at about the same time, 1969 and after, that a general disillusionment with politics had begun to erode the energy and commitment of those active in protest movements. Police and army "shoot-ups" of marches and political headquarters, the obvious presence of agent provocateurs, and the arrest and trial of activists for alleged crimes led to a general cooling of the New Left activity.[6] The weariness that many had come to feel in the endless rounds of meetings, rallies, and unsuccessful election campaigns had also taken its toll. It was during this period that many in the United States dropped out of political activity and began a certain kind of sociotechnical tinkering; roof gardens, solar collectors, and windmills became a focus of community action. A politics that had become too dangerous or depressing when faced with a showdown with real power could reestablish itself as a quest to reform "the system" through social and technical invention.[7]

This transition is clearly reflected in the contents of two catalogs of American radicalism published late in the decade: *Movement toward a New America*, edited by Mitchell Goodman, and *The Whole Earth Catalog*, edited by Stewart Brand.[8] Goodman's volume reprints hundreds of handbills, pamphlets, articles, and news stories that reflect the political activities of the 1960s. Rent strikes, the Black Panthers, SDS, antiwar demonstrations, radical feminists, and the like are described in the actors' own words and in the writing of sympathetic observers. The sense of a spirited, many-faceted nationwide movement challenging the oppressive hold of an established order is everywhere present. Published in 1970, the book was actually a last gasp. By the time it became possible to take a snapshot of all the different themes and actors, the movement had already fallen into decline. While *Movement toward a New America* offers itself as a handbook for people just getting underway, we can now see it as a scrapbook for those who were just getting out.

Although actually published before the Goodman collection, *The Whole Earth Catalog* expresses a tendency gaining momentum rather than withering away. Its vision is that of a groovy spiritual and material culture in which one's state of being is expressed in higher states of consciousness and well-selected tools. Citizenship in the American nation had at that time, 1968–69, evidently become an onerous burden for some. The Weathermen denounced the U.S.A. and declared that they were actually soldiers in the revolutionary armed forces of the Third World.[9] In contrast Stewart Brand's book consoles its readers with the idea that they are citizens of the planet Earth and its global systems: hippie environmentalist spacemen. Here the obsession with technology, especially *good* technology, a theme almost totally absent from Goodman's compendium, enters very strongly. In its statement of purpose, the *Catalog* announces that "a realm of intimate, personal power is developing—power of the individual to conduct his own education, find his own inspiration, shape his own environment, and share his adventure with whoever is interested." It explains that an item is listed "if it is deemed: (1) useful as a tool; (2) relevant to independent education; (3) high quality or low cost; (4) easily available by mail."[10]

If a group of people are moving back to the country and need a good sturdy cider press, *The Whole Earth Catalog* will tell them where to purchase one. Whole categories of small (and sometimes large) producers are brought in touch with a new generation whose "lifestyle" and habits of consumption have begun to change. *The Whole Earth Catalog* assumes that throngs of people will be moving off into small, humanly nurturing, economically self-sustaining communities that fit into a new complex world system destined to save the earth from the destruction of overindustrialization. In this vision, choices about the right technologies—both useful old gadgets and ingenious new tools—are crucial; choices about political matters are not. Preferring brief, enthusiastic descriptions of items for sale, *Whole Earth* avoids essays on topics of social, political, or even environmental concern. Evidently, the catalog-browsing consciousness suited to the new age is not one that wants to be bothered by well-reasoned arguments.

The Movement toward a New America and *The Whole Earth Catalog* express different sides of an era that had reached its crossroads. Perhaps the most telling comparison of the two works today would be that whereas the majority of addresses in Brand's volume identify producers and products still very much in business, the addresses of organizations and publications listed by Goodman are the stuff of dead letters.

The growing fascination with technology among those involved in movements of the 1960s was nourished by the writings of well-known European and American intellectuals. Starting from very different points of view, Lewis Mumford, Paul Goodman, Herbert Marcuse, Theodore Roszak, and Jacques Ellul carried the subject of modern technics and the technical mentality to the foreground of social criticism.[11] Marcuse's *One Dimensional Man* (1964) portrayed both capitalist and socialist societies as components of a vast, repressive technological civilization that was bringing every aspect of humanity under its control. Mumford's *The Myth of the Machine: The Pentagon of Power* (1970), a pessimistic conclusion of a lifetime of commentary on material culture, judged the promise of modern technics betrayed by the destructiveness of authoritarian megatechnics and the spiritual hollowness of expertise. Ellul's *Technological Society* (1964) provided its readers with an extreme statement in the same vein, arguing that every aspect of twentieth-century life— economics, politics, symbolic culture, individual psychology, and so forth—had fallen under the domination of *la technique*. Such books were read and widely talked about by those who found more orthodox modes of social analysis inadequate to express the troubles they saw in the modern world. Either there was something abominable in modern artifice itself, a position strongly argued by Jacques Ellul, or the particular varieties of artifice most commonly built in the modern age were simply the wrong ones, wrong in the sense that they generated destruction so vast as to undermine the very benefits of technological productivity. It was the latter belief, however reasonable or unreasonable it may be, that eventually helped spawn the idea that an alternative or appropriate technology was something that one could expect to discover or invent.

Of course, the misgivings about the direction of modern technological society did not begin with the movements or writings of the last twenty years.[12] Significant signs of questioning and opposition arose in the early nineteenth century and have reappeared in various manifestations each decade since. Thomas Carlyle's "Signs of the Times" (1829), for example, already contains many of the themes central to mid-twentieth-century criticisms of technology: awe at man's domination of nature, unease at the disruption of tradition, disgust with the regimentation of workers, rage at the injustices of the industrial economy, and anxiety at the loss of a moral center in the face of technical advance. "Men are grown mechanical in head and heart, as well as in hand," Carlyle exclaims. "They have lost faith in individual endeavour, and in natural force, of any kind. Not for internal perfection, but for external combinations and arrangements, for institutions, constitutions—for Mechanism of one sort or other, do they hope and struggle. Their whole efforts, attachments, opinions, turn on mechanism, and are of a mechanical character."[13]

The tendency to obliterate distinctions between technology and other social phenomena, a conceptual expedient characteristic of Ellul, Marcuse, and Roszak, is fully present in Carlyle's version of the story. In search of a comprehensive account of the malady and a deep-seeking diagnosis of its cause, writers in this tradition have more often than not arrived at conclusions that deny the possibility of practical remedy. Specific ills in industrial civilization are said to be rooted very deeply in human aggressiveness, the machine mentality, the essence of Western rational thought, the workings of the second law of thermodynamics, or in some similarly intractable source. Thus, over many generations, writings critical of technological society have been almost totally barren of suggested cures for the host of difficulties they so describe. Their lament is clear enough; their sense of outrage vivid. But the eloquence of criticism (perhaps this is a property of criticism) is matched by a poverty of practice. Since technology is the sphere in which "it works" is paramount, an uneasiness with problem-solving seems to undermine the very foundation upon which nontechnical thinkers base their credentials to talk about technical matters at all.

In summary, the basic themes, mode of analysis, central insights, and distinctive shortcomings of contemporary writings on technological society have a deep resonance with a tradition of theory and criticism that reaches back to the earliest stages of industrialism. Although the inadequacies of this way of thinking are manifest, it is also true that the litany of hopes and fears we find here lies at the center of much that is most lively in the social thought of the nineteenth and twentieth centuries. Hopes that one might find an appropriate technology arise at a time when the concepts and questions of this tradition have been renewed, restated for a new generation, and popularized for a mass audience. In the late 1960s and early 1970s, however, the sense of futility and despair so often characteristic of speculative writings on technology is suddenly replaced by an extravagant optimism. Thus we find the philosophers again and again pondering the issue of what the technology of an emancipated society would look like. They begin wondering aloud: Who might take up the tasks of realizing such a vision? Upon what occasion might their opportunity come?

Herbert Marcuse's *An Essay on Liberation* (1969), for example, offers an unabashedly utopian response to the dismal situations sketched in *One-Dimensional Man* (1964). He writes, "Freedom indeed depends largely on technical progress, on the advancement of science. But this fact easily obscures the essential precondition: in order to become vehicles of freedom, science and technology would have to change their present direction and goals; they would have to be reconstructed in accord with a new sensibility—the demands of the life instincts. Then one could speak of a technology of liberation, product of a scientific imagination free to project and design the forms of a human universe without exploitation and toil."[14] He is aware that such steps are possible "only after the historical break in the continuum of domination"[15] and tries to spell out the conditions of moral decay and economic and political disorder that could undermine and eventually topple the repressive society. Impetus for the change will come, Marcuse expects, from a combination of traditional working-class grievances, middle-class rebellion at conditions of surplus repression, and the quest for freedom and social justice among Third World peoples and the minorities of industrialized nations. The essay envisions "not the arrest or reduction of technical progress, but the elimination of those features which perpetuate man's subjection of the apparatus and the intensification of the struggle for existence."[16] In his emphasis upon a social reconstruction that includes political judgments about technological design features, Marcuse begins building a bridge between the themes of Frankfurt School critical theory and the possibility of an alternative technology.

In a move intellectually comparable to Marcuse's, Theodore Roszak's *Where the Wasteland Ends* (1972) concludes with a description of an alternative society, the "visionary commonwealth," that could overcome the destructive tendencies of technocratic civilization. Roszak's previous work, *The Making of the Counter Culture* (1969), focused upon the pervasive sickness of modern militarism, urbanism, consumerism, bureaucracy, and the technocratic mentality but ended with a song of praise to shamanism and the call to transcend the subject/object split.[17] It was not much of a program for action. The later book, however, matches its pleas for spiritual enlightenment with speculative proposals for decentralization, deurbanization, the creation of economically self-sufficient communities, and innovation toward a broad range of alternative technical and social forms. Calling attention to the history of theory and practical experiments in communitarianism and anarchist socialism, Roszak imagines in his list of proposals "the proper mix of handicraft labor, intermediate technologies, and necessarily heavy industry . . . the revitalization of work as a self-determining, non-exploitative activity . . . a new economics elaborated out of kinship, friendship and co-operation . . . the regionalization and grass roots control of transport and communication . . . non-bureaucratized, user-developed, user-administered services." Roszak is aware that in a world like the present one, such lovely plans are likely to sound foolish. "I can think of forty reasons," he explains, "why none of (these) projects can possibly succeed and forty different tones of wry cynicism in which to express my well-documented doubts. But I also know that it is more humanly beautiful to risk failure seeking for the hidden springs than to resign to the futurelessness of the wasteland."[19]

Roszak's comment is a revealing one. For many are apt to conclude that those who have become interested in appropriate technology or decentralization are simply naive about the way things work. If they knew the facts, they would respond much differently. Either they would recognize the existence of overwhelmingly powerful economic and political institutions in our time and adapt their strategies to achieve realistic reforms, or they would face their oppressors head on in a struggle against capitalism, militarism, and technocracy.

But certainly writers like Roszak, Marcuse, and Mumford are not unaware of the existence of multinational corporations, the Pentagon, the CIA, AT&T and other bodies that wield power in our time.

It is not naiveté they show toward such institutions but rather total contempt informed by a sense of powerlessness. If they are to avoid the cynicism and gloom toward which their own best thinking carries them, they must perform a highwire act along very slender threads of hope.

Appropriate technology in industrialized countries, I believe, begins on much the same tightrope. Stemming from the decline of radical politics and from an obvious next step in the critique of technological society, its first purpose is not to produce energy from renewable resources but to generate hope from the winds of despair.

THE NEW AGE: THEMES AND CONTRADICTIONS

But the puzzle arises: Why should *technology* be the focus here? Why not an alternative economics, for example? Indeed, in Europe and America during the late 1960s and early 1970s, many young politically oriented intellectuals returned to Marxian economics as a way of deepening their understanding of society.[20] The founding of the Union for Radical Political Economics in the United States comes at about the same time as the founding of the New Alchemy Institute and reflects a similar sense of intellectual possibilities. At stake in both radical political economy and appropriate technology is an attempt to interpret the realm of material culture in a fresh and revealing light. In both the perspective on this subject called "economics" and the approach called "technology," long-standing, stultifying orthodoxies had been silent on many of the questions that seemed most urgent. Thus, the neo-Keynesian economics taught in universities took the existing relations of production as given and paid attention to increasingly refined ways, including econometrics, of keeping "the economy" (*the* economy?) in tune. For those who had become convinced that there were fundamental questions of social structure to be addressed in any science of economics worthy of the name, the academic orthodoxy simply would not do.

If political economics has become a field ripe for imaginative research and theory, the sphere of things called "technology" is also a realm virtually wide open to fresh interpretations. Long considered something too ordinary or too well-known to permit lively speculative inquiry, technology was for generations ruled by a rigidly frozen orthodoxy that recognized little of human interest in the subject other than the ingenuity of problem-solving. If there were any ethical, aesthetic, political, or metaphysical issues relevant to technology as a field of human enterprise (and surely there were always many), they were almost never addressed by engineers or by other technical professionals. Questions about the *meaning* of the techniques, apparatus, and systems that characterize modern technology were, by some implicit understanding of those best situated to address them, virtually closed to the inquiring mind. Only on rare occasions during the twentieth century would respectable technologists bother to raise the most obvious social and ethical problems surrounding their work. Even then their thoughts tended to be platitudinous and self-serving.[21] It was left, therefore, to intrepid outsiders like Lewis Mumford to raise important issues that prudent men avoided.

As an interest in appropriate technology spread during the mid-1970s, many technical specialists began to investigate novel styles of interpretation and action in science and engineering. As the rosters of such organizations as the Brace Research Institute, Faralones Institute, Goddard College, the Intermediate Technology Development Group, and the State of California Office of Appropriate Technology clearly show, a good number of those engaged in opening these boundaries have excellent backgrounds in biology, physics, electrical engineering, architecture, and other scientific and technical fields. They presumably know what "the hard stuff" is and what the dominant orthodoxy has to say about how one properly follows one's profession in the "real world." But recognizing a need to set their knowledge and skills to fundamentally new purposes—purposes outside those normally recognized by the corporations, foundations, government, and universities—they have sought to redefine the context in which scientific research and technological innovation can take place. True, there are also many who see appropriate technology as a convenient label to attract government and foundation grants so that they can continue doing what they have always been doing. But what interests us here are not the inevitable phonies in any field but those who, despite their difficulties, are making an honest effort to do something importantly new.

To take one of several possible examples, in both form and content *The Journal of the New Alchemists* reflects the desire to do science and technical invention in a different key. Contained in its volumes are reports on aquacultural and agricultural experiments, diagrams of solar and wind devices, tables of data on research underway, and other features that reflect the canons of modern scientific methodology. Interspersed with these, however, are poems, philosophical essays, drawings, and highly personal accounts of the experience of the institute's members. Introducing the article by aquaculturalist Bill McLarney, "Studies of the Ecology of the Characid Fish *Brycon guatemalensis* in the Rio Tirimbina, Heredia Province, Costa Rica with Special Reference to Its Suitability for Culture as a Food Fish," the editor comments on McLarney's research procedures: "Clad in sun hat, fisherman's vest, shorts, and running shoes he pursued his quarry through the water with Chaplinesque élan. In both his science and his writing his standards of excellence are irreproachable, but being around the work actually in process is to be part of an on-going comedy."[22]

To present one's endeavors in this fashion, of course, deliberately risks the credibility of the New Alchemy Institute within a society of scientists and engineers. The accepted form of "objectivity" in scientific and technical reports nowadays (one can also include books and articles in social science) requires that the prose read as if there were no person in the room when the writing took place. While the New Alchemists recognize the standards of experimentation and evidence prevalent in modern research, they do not adhere to notions of objectivity that conceal the social purposes of science. Their own aims are explicitly stated on the inside cover of each volume. "We seek solutions that can be used by individuals or small groups who are trying to create a greener, kinder world. It is our belief that ecological and social transformations must take place at the lowest functional levels of society if people are to direct their course toward a saner tomorrow." According to a story frequently told by New Alchemy president John Todd, the institute formed when a group of young biologists finally grew tired of attending prestigious international conferences at which calamity for the planet through pollution, overpopulation, and other environmental dangers was blithely predicted by experts whose research commitments seemed to preclude any attention to how such disasters might be averted.

The existence of a growing number of research institutes, small firms, government agencies, universities, philanthropic organizations, and independent individuals that claim to be doing appropriate technology is not a sign that such people share a common philosophy. In Amory Lovins' view of the matter, for example, appropriate technology is fully compatible with capitalism, albeit a capitalist system reorganized to provide investment funds for the "soft energy path."[23] David Dickson, a British writer on science and technology, describes the possibilities from a Marxist perspective and believes that "alternative technology" represents a new chapter in the history of socialism.[24] Murray Bookchin, veteran political activist in the American Left and teacher of appropriate technology at Goddard College, joins many spokesmen—including Theodore Roszak and Ivan Illich—in an interpretation that finds in the movement a rebirth of communitarian anarchism.[25] Not to be outdone, the National Academy of Sciences commissioned economist Richard Eckaus to do a study that evaluates appropriate technology through the dry and predictable criteria of the academic establishment: economic growth, transfer of technology, maximization of output, "political development," and the like.[26] Indeed, as people from different backgrounds and persuasions try to move from staid orthodoxies toward new understandings, one would expect that there would be a flurry of attempted refoundings, not all of them compatible. What we are witnessing here may be less a genuine paradigm shift than a grinding of ideological gears.

In the inevitable attempt to overcome uncertainty and to achieve order in the face of obvious conceptual chaos, a number of observers have tried to argue that "appropriate" or "alternative" or "soft" technology is a uniform movement with a logically coherent set of characteristics. More than that, of course, many seem to believe that this set of characteristics constitutes a univocal ideal, a picture of a world appreciably better than the one we now inhabit. In his book *The Politics of Alternative Technology*, David Dickson quotes with approval a list of thirty-five criteria for "soft technology" prepared by Robin Clarke of Biotechnic Research and Development in England. Dickson observes that they "form a coherent system that can only be interpreted as a whole, and loses much of its sense when reduced to fragmented components."[27] Thus, "soft technologies" are characterized as "ecologically sound, small energy input, low or no pollution rate, reversible materials and energy sources only, functional for all time, craft industry, low specialization, . . . integration with nature, democratic politics, technical boundaries set by nature, local bartering, compatible with local culture, safeguards against misuse, dependent on well-being of other species, innovation regulated by need, steady-state economy, labour intensive, . . . decentralist, general efficiency increases with smallness, operating modes understandable to all," and so forth.[28] In the other list, "hard technology" is described as "ecologically unsound, large energy input, high pollution rate, non-reversible use of materials and energy sources, functional or limited time only, mass production, high specialization, . . . alienation from nature, consensus politics, technical boundaries set by wealth, world-wide trade, destructive of local culture, technology liable to misuse, highly destructive to other species, innovation regulated by profit and war, growth-oriented economy, capital intensive, . . . centralist, general efficiency increases with size, operating modes too complicated for general comprehension," and other such qualities.[29]

The lists are astounding! Criteria for hard technology read like a complete index to the sordid perils of modernity. In contrast, the criteria for soft technology provide nothing less than a list to Santa Claus asking for paradise for Christmas. I would not deny the importance of expressing one's best dreams of and entertaining visions of a better world. Political and social thought that excludes such imagination is apt to become sterile and resigned to accept the failures in our present institutions. But the portrait sketched by Clarke and approved by Dickson woefully confuses the role of both social criticism and utopian thinking.

Nothing in Western philosophy or, for that matter, nothing in human experience indicates to us that we can arrange the good and the bad in simple, noncontradictory lists. If one pursues an idea of justice far enough, for example, it may well begin to conflict with one's own cherished notion of freedom. Neither the intricacies of theory nor the history of practice give us any reason to believe that Clarke and Dickson's desire to harmonize community, self-sufficiency, safety, diversity, democracy, and efficiency is at all reasonable. It is not merely that such wishes are not feasible in practice, that certain facts of the world stand as barriers to their realization. More importantly, the selection of concepts upon which such a vision of good technology rests is fraught with incompatibles. It is not obvious, for example, that decentralized production is necessarily environmentally sound; that labor-intensive technologies provide "work undertaken primarily for satisfaction"; that small-scale communities encourage social diversity. If one wanted to pick an anthem for appropriate technology, it would be wise for the time being to avoid rousing victory marches and proceed directly to Charles Ives's "Unanswered Question."

Roughly speaking, of course, we can identify the kinds of reasons that people have given for getting started in appropriate technology. Most of these reasons are present on Clarke's soft-technology list. Many people have become concerned about what they perceive to be "the limits to growth"; many would like to respond to the coming shortage of fossil fuels with energy systems based on renewable resources; many worry about how to solve the world population and hunger problems; many would like to see local control over political, social, technical, and economic decisions; and so forth. But there is something strangely incomplete about the group of reasons normally given. The astonishing fact that must be accounted for is why so many members of the North American and European middle class, arguably those *best served* by the technological society as it now stands, have become fascinated with something they call "appropriate technology."

Among the places where the symbolism and underlying motives of the movement are most clearly visible are the fairs and expositions that now celebrate the coming of the so-called New Age. In much the same way that the great world's fairs of the nineteenth and twentieth centuries offered an opportunity to witness our civilization's grandest (and often least realistic) hopes for science and industrial technology, expositions of the sort held recently in Los Angeles, San Francisco, Vancouver, Boston, and other North American cities tell us a great deal about the images that surround appropriate technology. Of course, it would be a mistake to take much of it at face value. Seeing a working model of a composting toilet in a convention hall offers you no better sense of what composting toilets might mean for society than seeing a model automobile assembly line at a world's fair forty years ago gave visitors a valid sense of the social significance of either the auto or the assembly line. Nevertheless, as a way of holding one's finger to the pulse of a certain range of popular developments, such fairs provide a good opportunity.

One such fair that I was able to visit and study in person was the New Earth Exposition held in Boston in early May 1978. The theme of the exposition was "Living Lightly on the Earth." Its advertising promised it would be "a showcase for environmentally creative individuals and businesses demonstrating viable alternatives in the field of: Energy, Personal Growth, Food, Transportation, Shelters, Gardening, Recycling, Wilderness Skills, Ecology, and Publishing." On the floor of the huge Commonwealth Pier exhibition hall were hundreds of small (and some not so small) organizations, showing their wares and passing out literature. I was astonished at the sheer number of groups represented and surprised at how far they had come in developing their goods and services. In the category of wood-burning stoves alone there were no fewer than a dozen entrants. Choosing a wood-burning stove has become as difficult as selecting a new car.

In the general run of items and ideas displayed at the exposition, those most clearly fitting the category "appropriate technology"—solar collectors, owner-built homes, energy-saving devices, windmills, wood stoves, solar greenhouses, and other kinds of new-fangled and old-fashioned hardware—were, oddly enough, also those that most readily fit the model of ordinary American respectability. Along with their explanations about the environmental benefits to be gained from using these devices, the exhibitors stressed the virtues of sound engineering, thrift, and good business sense. Thus, at one and the same time the salesmen appealed to the very conservative as well as radical understandings of what an intelligent person ought to be doing these days.

Other parts of the show catered to the visitors' tendencies toward escapist consumerism and spiritual self-indulgence. A large cluster of exhibits advertised new toys of the young, professional middle class: jogging shoes, racing bicycles, hot tubs, backpacks, tents, hang gliders, and so forth. The traditional American notion that freedom means something like the ability "to hit the road whenever I feel like it" was fully represented with hiking equipment and sleeping bags replacing the automobile as vehicles for realizing this fantasy.

Promoters of alternative health care measures were also conspicuously present. An amazingly large number of booths presented the wonders of New Age medicine—acupuncture, Shiatsu massage, macrovitamin therapy, herbal remedies, iridology, and other such practices—a new growth industry in which many now invest time and money that would otherwise have gone to the American medical establishment. In an adjacent field of concern, advocates of a rapidly proliferating array of spiritual disciplines sought converts from among passers by—yoga, meditation, Arica, the sufis, t'ai chi, pyramid power, and several kinds of spiritual communities. The market was open for anyone buying. But as one might have predicted, explicitly political causes were pretty much absent from the event. Environmental activists and organizations opposing nuclear energy or trying to save the whales were about the extent of the groups that bothered to secure the fairly easy to obtain exhibitors' invitations. Evidently, the New Earth in its New Age will be able to get along without politics as such.

What is going on here? What do expressions of this kind tell us about the climate of opinion in which appropriate technology has arisen? Beyond an honest concern to preserve the natural environment, beyond even the wish to rediscover community, personal autonomy or both, there is a significant obsession that emerges from the symbolism of fairs like Boston's New Earth Exposition. Underlying the vast majority of exhibits and not too far below the surface of their claims was the simple need to overcome tension—the stresses and strains of modern civilization. Many people now thoroughly enmeshed in the demanding business of keeping our technological society running have begun to hedge their bets. Through bicycling, transcendental meditation, running, organic food, massage, backpacking, yoga, and a host of ingenious means they seek the relaxation and peace of mind that everyday life cannot provide. Yes, technologies of a certain kind are required. But they are not inventions like solar hot water heaters. Wanted instead are devices and techniques that can alleviate the kinds of pressures that normally befall professionals toiling in banks, insurance companies, and bureaucracies. Rather than attempt to change the structures that vex them, young Americans growing older have settled on the quest for exquisite palliatives. If the 1960s announced, "Let's see if we can change this society," the 1970s proclaimed, "Let's get out of this skyscraper and go jogging!"

Contained in the frantic quest for health, vitality, and self-realization is a growing distrust of the goods and services of American abundance as well as a distrust of the professionals and organizations that provide them. Who needs doctors? Do your own health care. Who needs architects and contractors? Build your own home. Who needs the utilities? Generate your own energy. The desire for "self-sufficiency," long regarded as a virtue in Western culture, is here clouded by a profound resentment. Cramped by inflation, pressured in unsatisfying jobs, fearful of cancer and heart attack, bored with the products of the consumer economy, and unable to escape to the once glorious dream of happiness in suburbia, a substantial number of those who have "made it" in modern society are now restless.

THE BETTER MOUSETRAP

The failure of the dream of American materialism would be easier to accept if it were merely the case that a blight on individual souls accompanied the manufacture of an otherwise perfect artificial paradise. Then one could be "born again" through Christianity or yoga and feel better as one drove down the freeway or strolled through the shopping mall. But as many citizens of the technological society have begun to notice, hollowness and corruption afflict *the very objects and institutions of material culture itself.* Because the automobiles, appliances, prepared foods, and diverse technological products and services have all too often been designed and built according to degenerate specifications, the good life begins to look like a colossal pile of junk.

In this understanding of things, what is at stake in appropriate technology is a war against qualities that William Morris used to call "the shoddy." An amusing and insightful spokesman for this position in our own time is Victor Papanek, chairman of design at the Kansas City Art Institute. Papanek's *How Things Don't Work* offers seemingly endless examples of deplorable screw-ups in the design and engineering of our everyday tools and consumer goods. "Between 1966 and February 1975 the automobile industry recalled

45,700,000 automobiles for inspection or repair," he observes to indicate how bad things have gotten.[30] Beneath a photograph of two commercial cheese graters he notes, "On the left, the inexpensive, efficient, and nearly indestructible one which will work left- or right-handed. On the right, the "improved model, which is right-handed only and, after some months of use, grinds its own plastic coating into the food."[31]

In a haphazard rather than systematic analysis, *How Things Don't Work* locates a wide range of causes of shoddy industrial design and manufacture. Unwise cost-cutting in the name of profit diminishes the quality of many goods. Alienated assembly-line workers take secret revenge upon the automobiles and other products that pass by them on conveyor belts. The effects of advertising and showcase packaging decrease the usefulness of previously useful items. Senseless so-called improvements in the name of "convenience," "ease of handling," and "customer comfort" drive up the cost but degrade the utility of home tools and appliances. As one reads Papanek's observations, it seems that what he has succeeded in doing is to translate the criticisms of Mumford, Ellul, Marcuse, and Illich into a set of perverse criteria for technical invention and innovation. And they seem only slightly less bizarre when we notice that they are many of the criteria upon which our industrial economy is actually founded.

Papanek's response to all of this is an interesting one. His prescription not only reveals his own understanding of what is to be done but also reflects a model of social change implicit in the writings and projects of appropriate technologists of all persuasions. "Massive problems faced by workers and users need innovative remedies," he explains, and he offers a brief survey of frequently proposed nostrums. "There is the capitalist approach (make it bigger), the technocratic one (make it better), the 'revolutionary' solution (portray the problem as an example of an exploitative system) and the preindustrial romantic fallacy (don't use it; maybe it will go away by itself). We propose a fifth alternative response: *Let's invent a different answer.*"[32]

Donning the mantle of the old-fashioned tinker/designer/crackpot inventor, Papanek describes an array of clever techniques and devices that he and his colleagues around the world have created. A pedal-driven lawn mower, fold-up bathtub, fold-up bicycle, spillproof scooter, do-it-yourself emergency service vehicle, and other contraptions "fitted to human scale" are depicted in drawings and photographs. Innovations in hardware, he points out, must be matched by innovations in patterns of ownership and use. "Few tools in our society are designed for communal (or shared) ownership. If they were designed for sharing, rather than for individual use, we believe they would change structurally, mechanically, and in material composition. When reel-type push mowers were still in use, for example, they were more frequently borrowed or shared than today's electric or gasoline-driven mowers for the simple reason that they were robust, uncomplicated, and difficult to break."[33]

Papanek's guiding maxim, "Let's invent a different answer," echoes a familiar apothegm usually attributed to Ralph Waldo Emerson (although it is not certain that he said exactly these words): "If a man can write a better book, preach a better sermon, or make a better mousetrap than his neighbor, though he builds his house in the woods the world will make a beaten path to his door."[34] This is, of course, the traditional American notion about how inventions change the world. A good idea or invention will catch on. In fact, as Emerson evidently wished to say, if one's scheme is good enough, there will be no stopping it. Ingenuity creates its own demand.

As scholars Arthur Bestor and Dolores Hayden have observed, nineteenth-century American utopians held much the same conviction with regard to how their experiments might eventually transform all of society. The utopians believed that their technical and social inventions would have a strong appeal to an age undergoing rapid change. Communities like those at New Harmony and Oneida saw themselves perfecting what Bestor calls "patent office models" of the good life.[35] In the same way that ordinary people would eagerly accept new improvements in farm machinery if a convincing demonstration were given them, so would they be willing to embrace the principles and devices of utopia if a successful working model could be built and maintained somewhere in the world.

Insofar as they have any coherent idea of how their labors will change the world, appropriate technologists appear to entertain similar views. A person builds a solar house or puts up a windmill, not only because he or she finds it personally agreeable, but because the structure will serve as a working model to inspire others to do likewise. If enough folks build or retrofit for renewable energy, so the unspoken assumption goes, there will be no need for the nation to build a system of nuclear power plants. Thus, individuals will, in effect, vote on the shape of the future through the technological choices they make today. This notion of social change provides the underlying rationale for amazing emphasis on do-it-yourself manuals, catalogues, demonstration sites, information sharing, and "networking" that now characterizes appropriate technology in North America and Europe.[36] Once people discover what is available to them, they will send away for the blueprints and build the better mousetrap themselves. As successful grassroots experiments proliferate, those involved in similar projects will get in touch with one another and begin to form little communities, slowly reshaping society through a growing aggregation of small-scale social and technical innovations. Radical social change will catch on like smoke detectors.

It is not difficult to find fault with such ideas. The same misgivings that Marx and Engels had about the anarchists and utopian socialists of their day apply just as well to the appropriate technologists of our time. As they sketch their lovely plans, they show a remarkable unwillingness to face the facts of economic and political power. They seem, furthermore, unwilling to engage in the kind of deep-seeking economic and institutional analysis that would show the futility of their schemes. The appropriate-technologists' hopes for a decentralized polity that arrives through massive conversion to solar energy is already being undermined by the attempts of American corporations and public bureaucracies to shape developments in this field along lines that correspond to existing institutional patterns. Large corporations in electronics, metals, oil, and glass have already begun to invest heavily in developing and marketing solar products. Federal funding in energy research shows a marked preference for projects on photovoltaics and other high-tech versions of the "soft energy" path. The story of appropriate technology may well turn out to be one in which the inertia of existing sociotechnical systems overwhelms and absorbs a variety of social idealism that set out to provide modern technology with new designs and directions.[37]

Indeed, building a better mousetrap removes none of the formidable barriers along the world's path to a new door. But having noticed this fact, one must also remember that the most important characteristic of all utopian radicalism past and present is its willingness to tackle projects despite reality's mocking glare. In this respect, a crucial but often overlooked intention of nineteenth-century utopians and the more ambitious of today's appropriate technologists is to claim the very language of interpretation about technical matters as a domain of moral and political discourse. Their conceptions of social change, so woefully inadequate from the view of practical immediacies, can better be seen as attempts to borrow on the credit of popular notions of "progress" and "obsolescence" and to give them a new significance. The task for radical utopians thus becomes that of trying to broaden the meaning of such categories as "efficiency," "rationality," "productivity," "cost" and "benefit." By suggesting moral and political contexts in which such terms ought to find their significance, it may be possible to revise the grounds upon which technical judgments are made. Familiar criteria such as the norm of efficiency might eventually be seen in a new light. Whole new consid-erations, for example "the interconnectedness of things," might be added to the range of circumstances that an intelligent person would have to take into account when making a decision.

Strongly at issue in the writings and practical projects of appro-priate technology, therefore, are a variety of arguments about the need to limit the effects of human intervention on ecosystems, about the quality of energy sources as related to end use, about "second law efficiencies," and about the ways in which technologies both mirror and enforce fundamental social choices. Taken in sum, such arguments express an orientation toward natural and artificial sys-tems very much different from the attitudes of cavalier exploitation that have governed "progress" for the past two centuries. Those who have taken up these ideas show an active desire to transcend a condi-tion long lamented by philosophers and poets—the destructive nar-rowness of the standard categories of analysis, of the quantitative methodologies and practices of decisionmaking that have shaped the institutions, including technological institutions, of modern times. They recognize that in the early stages at least, the better mousetrap must succeed as conceptual invention.

In this regard, however, an important weakness in the work of appropriate technologists so far has been the lack of any serious attention to the history of modern technology. Presumably, if the idea of an appropriate technology makes sense, one ought to be able to discover points at which developments in a given field took, in retrospect, an unfortunate turn, points at which the choices made produced an undesirable technical ensemble from the possibilities then available. One might, for example, look at the range of discover-ies, inventions, industries, and large-scale systems described in works like Alfred D. Chandler's *The Visible Hand* to understand which paths in modern technics were actually followed.[38] One might then attempt to come to terms with such questions as: Why did develop-ments proceed as they did? Were there any realistic alternatives? Why weren't those alternatives selected? How, if at all, is it possible to reclaim the sense of choice within a given set of technical possibilities such that one might shape the outcome of that choice according to insights that our own history now makes available? In their investi-gations on energy and agriculture some appropriate technologists have begun asking such questions to a limited degree. But by and large most of those active in this area have been willing to proceed as if history simply did not matter.

It is true of course that in a situation like the present one any attempt to achieve fundamental change in the order of artificial systems will seem inherently self-contradictory. A man builds a beautifully designed solar home in Vermont and is proud of its energy savings. He then flies in a jumbo jet to an environmental conference in London to report his success. The man would do well to ask himself: Why am I saving energy and for whom?

Appropriate technology finds its work in a world of bizarre moral inversions that constitute the source of twentieth-century perplexities. While millions of people around the globe go without adequate food, shelter, or work, nearly $400 billion is spent each year on increasingly lethal armaments that are said to secure the "peace." Nations ignore or revise downward their standards of environmental quality because pollution controls are incompatible with a "sound economy." In situations like these, we become used to accepting deplorable conditions as normal. To be realistic and to get things done almost requires that a person become the enemy of a free humanity and a healthy planet.

The condition we face is very much like that described in Bertold Brecht's beautiful little play, *The Exception and the Rule.* On Brecht's stage, a handful of characters wander through a pattern of actions that reveal a moral universe turned on its head. What is good is made to appear evil; justice and injustice trade places. A coolie attempts a good deed and is killed by his merchant employer, who sees the coolie's gesture as a threat from a class enemy. The murderer placed on trial is acquitted in a judgment that finds his behavior perfectly reasonable under the circumstances. Brecht is never one to let his audience miss the point. At the beginning of the play his actors exclaim:

> Inquire if a thing be necessary
> Especially if it is common
> We particularly ask you—
> When a thing continually occurs—
> Not on that account to find it natural
> Let nothing be called natural
> In an age of bloody confusion
> Ordered disorder, planned caprice,
> And dehumanized humanity, lest all things
> Be held unalterable![39]

NOTES TO CHAPTER TWO

1. E.F. Schumacher, *Small Is Beautiful: Economics as If People Mattered* (New York: Harper & Row, 1973).

2. Surveys of organizations engaged in appropriate technology research and development include: *Appropriate Technology—A Directory of Activities and Projects*, Integrative Design Associates (Washington, D.C.: U.S. Government

Printing Office, 1977); *A Handbook on Appropriate Technology*, Canadian Hunger Foundation and Brace Research Institute (Ottawa: Canadian Hunger Foundation, 1976); Ken Darrow and Rick Pam, *Appropriate Technology Sourcebook* (Stanford, Calif.: Volunteers in Asia, 1976). The best general introduction to the topic is Nicolas Jéquier's essay, "The Major Policy Issues," in *Appropriate Technology: Problems and Promises*, ed. Nicolas Jéquier (Paris: Development Centre of the OECD, 1976), pp. 15-112.

3. See Richard Eckaus, *Appropriate Technologies for Developing Countries* (Washington, D.C.: National Academy of Sciences, 1977); and *Appropriate Technology in World Bank Activities*, The World Bank, July 19, 1976.

4. Schumacher, p. 31.

5. See, for example, Robert Owen, *A New View of Society and A Report to the County of Lanark*, ed. with an introduction by V.A.C. Gatrell (Baltimore: Penguin, 1970); Peter Kropotkin, *Fields, Factories and Workshops* (Boston: Houghton, Mifflin, 1899); Ralph Borsodi, *Flight from the City* (New York: Harper & Row, 1933); Peter van Dresser, *A Landscape for Humans* (Albuquerque: Biotechnic Press, 1972); Murray Bookchin, *The Spanish Anarchists: The Heroic Years 1868-1936* (New York: Free Life Editions, 1977).

6. An extremely insightful treatment of this period is contained in Jo Durden-Smith, *Who Killed George Jackson?* (New York: Knopf, 1976).

7. Perhaps the best expression of this tendency is to be found in Karl Hess, *Dear America* (New York: Morrow, 1975), chs. 8-10.

8. *The Movement toward a New America: The Beginnings of a Long Revolution*, assembled by Mitchell Goodman (New York: Knopf, 1970); *The Whole Earth Catalog*, ed. Stewart Brand and colleagues (Menlo Park, Calif.: Nowels, 1968).

9. See *Weatherman*, ed. Harold Jacobs (New York: Ramparts Press, 1970), for expressions of this view.

10. *The Whole Earth Catalog*, p. 1.

11. Lewis Mumford, *The Myth of The Machine: The Pentagon of Power* (New York: Harcourt Brace Jovanovich, 1970); Paul Goodman, *The New Reformation: Notes of a Neolithic Conservative* (New York: Random House, 1970); Herbert Marcuse, *One-Dimensional Man: Studies in the Ideology of Advanced Industrial Society* (Boston: Beacon Press, 1964); Theodore Roszak, *The Making of a Counter Culture; Reflections on Technocratic Society and Its Youthful Opposition* (Garden City, N.Y.: Doubleday, 1969); Jacques Ellul, *The Technological Society*, trans. John Wilkinson (New York: Knopf, 1964).

12. For an exposition of the central notions in this tradition see my *Autonomous Technology: Technics-out-of-Control as a Theme in Political Thought* (Cambridge, Mass.: MIT Press, 1977).

13. Thomas Carlyle, "Signs of the Times," in *Selected Works, Reminiscences and Letters*, ed. Julian Symons (Cambridge, Mass.: Harvard University Press, 1957), p. 25.

14. Herbert Marcuse, *An Essay on Liberation* (Boston: Beacon Press, 1969), p. 19.

15. Ibid.

16. Ibid., p. 90.

17. Roszak, ch. 6.

18. Theodore Roszak, *Where The Wasteland Ends: Politics and Transcendence in Postindustrial Society* (Garden City, N.Y.: Anchor Books, Doubleday, 1973), p. 396.

19. Ibid., p. 394.

20. See any issue of *The Review of Radical Political Economics* for the directions that Americans have taken here.

21. For a recent example see Samuel C. Florman, *The Existential Pleasures of Engineering* (New York: St. Martin's Press, 1976). An interesting controversy from an earlier period between critics of technological society and leading spokesmen in science and engineering can be found in *Whither Mankind: A Panorama of Modern Civilization*, ed. Charles A. Beard (New York: Longmans, 1928); and in *Toward Civilization*, ed. Charles A. Beard (New York: Longmans, 1930).

22. *The Journal of the New Alchemists* 1 (1973): 50.

23. Amory B. Lovins, *Soft Energy Paths: Toward a Durable Peace* (Cambridge, Mass.: Ballinger, 1977). See also Lovins' "A Neo-Capitalist Manifesto: Free Enterprise Can Finance Our Energy Future," *Politicks and Other Human Interests* 1, no. 2 (April 12, 1978): 15-18.

24. David Dickson, *The Politics of Alternative Technology* (New York: Universe Books, 1975).

25. Murray Bookchin, *Post-Scarcity Anarchism* (Berkeley, Calif.: Ramparts Press, 1971).

26. Eckaus, *Appropriate Technologies for Developing Countries*.

27. Dickson, *The Politics of Alternative Technology*, p. 102.

28. Ibid., pp. 103-104.

29. Ibid.

30. Victor Papanek and James Hennessey, *How Things Don't Work* (New York: Pantheon Books, 1977), p. 49.

31. Ibid., p. 119.

32. Ibid., xiii.

33. Ibid., p. 27.

34. The saying, attributed to an address Emerson gave, appears in Sarah S.B. Yule and Mary S. Keene, *Borrowings* (1889).

35. Arthur Eugene Bestor, Jr., *Backwoods Utopias: The Sectarian and Owenite Phases of Communitarian Socialism in America, 1663-1829* (Philadelphia: University of Pennsylvania Press, 1950). See also Dolores Hayden, *Seven American Utopias: The Architecture of Communitarian Socialism* (Cambridge, Mass.: MIT Press, 1976).

36. This emphasis is readily apparent in *Rainbook: Resource for Appropriate Technology* (New York: Schocken Books, 1977).

37. Typical of the attacks on appropriate technology from the left is Alexander Cockburn and James Ridgeway's "The Myth of Appropriate Technology," *Politicks and Other Human Interests* 1, no. 4 (December 6, 1977). "Technology," they argue, "whether alternative or intermediate, or even just big, can only exist as a reflection of the existing political and economic system. Appropriate technology, per se, won't change anything. A politics that may include

some aspects of smaller technology may lead to change—but the politics must come first" (p. 28).

38. Alfred D. Chandler, Jr., *The Visible Hand: The Managerial Revolution in American Business* (Cambridge, Mass.: The Belknap Press of Harvard University, 1977).

39. Bertold Brecht, "The Exception and the Rule," in *The Jewish Wife and Other Short Plays*, English versions by Eric Bentley (New York: Grove Press, 1965), p. 111.

Chapter Three

A Critique of the Concept of Appropriate Technology

Harvey Brooks

One of the difficulties in the discussion of appropriate technology (AT) is the variety of meanings attached to the concept by different proponents and critics. In consequence the exponents of different points of view often talk past one another and do not address the same issues.

One can distinguish two sets of themes or opposing positions in the discussion. In one set AT is regarded as an alternative technological system that is totally incompatible with existing technologies.[1] In this view not only the underdeveloped countries (LDCs) but the developed countries (DCs) must abandon the technological path taken by the West in the last century or so and start along a revolutionary new path in which the values and criteria alleged to be implicit in contemporary technology are explicitly and consciously rejected.

The opposite view is that AT is a romantic aberration, feasible only in special circumstances and not generalizable to large populations. Advocating it as a path for development of LDCs is simply a colonialist "put down," implying that the poor parts of the world cannot and should not aspire to the material standards of the industrialized countries.[2]

The proponents of AT retort that in fact current patterns of development paralleling historical economic development in the West cannot be extended to a world of 10 billion or more people, that AT is the only route by which the poor of the world can achieve human dignity, and the West must ultimately follow their path, not vice versa.

An intermediate position, which I shall espouse in this essay, is that AT and current technology are complementary rather than mutually exclusive, and that the potential benefits of both will be enhanced when they coexist. Indeed in many cases the criteria for AT do not stand up to analysis when the attempt is made to oppose or contrast them with present Western technology. Rather, there is a continuum of technologies with varying social characteristics, and the natural selection process by which the technological system evolves is likely to leave "ecological niches" for both AT and conventional technology in mutually supportive relationships. I will put forward reasons for my belief that the technological system is evolving toward greater diversity and pluralism, mirroring what is also happening with social characteristics in the most advanced industrial societies.

The second set of themes concerns the relative role of economic and social criteria in the social selection of technologies. One set of proponents of AT argues that it is more efficient even by traditional economic criteria and that modern technological trends are only viable because of massive intervention by governments to support and sustain them. In this view the recent evolution of the technological system is primarily the result of institutional inertia, supported by the capacity of those having a vested interest in conventional technology to register these interests effectively through the political process. In addition, many modern technologies such as nuclear power, chemical insecticides, interstate highways, and jet aircraft have derived or benefited from direct government research and development or subsidy.

A closely allied view is that AT, regardless of whether or not it is "efficient" by traditional economic criteria, is to be preferred because of the social values it promotes and with which it alone is compatible. These are values associated with such themes as "human scale," minimal ecological impact, the fostering of democratic participation and a sense of community, and, above all, compatibility with an egalitarian ethic.

The critics of AT, however, are basically skeptical of its claim to greater economic efficiency, no matter what the economic ground rules chosen. They argue that if AT were really "appropriate" in any meaningful sense, the free market would preferentially select it. They are skeptical that the institutional and political forces portrayed by AT proponents as artificially sustaining conventional technologies would be sufficient to override its market superiority if in fact AT were truly superior. Against those AT proponents who argue from social criteria, the critics reply that these criteria are either unrealistic or already met with current technology, that the free market gives the largest scope to individual choice and personal freedom, and that attempts to apply noneconomic criteria to the selection of technologies inevitably embody some form of totalitarianism in which the subtly elitist preferences of minorities are imposed on majorities through political methods.

With respect to the foregoing issue, I take the position that the distinction between economic and social criteria is not as sharp as implied, and that social criteria can be and are incorporated in the ground rules of the market, but that these social criteria are not as uniquely or obviously related to technical characteristics as AT proponents contend. And for exactly this reason what is called AT and conventional technology can coexist and fulfill a variety of social criteria while still permitting a high degree of personal choice. What I shall argue is that from the standpoint of social criteria technology may be regarded as a "black box." What is important is not what is inside the black box, but rather the "transparency" of the interface to the user.

APPROPRIATE TECHNOLOGY AND CONVENTIONAL TECHNOLOGY: MUTUALLY EXCLUSIVE OR SYMBIOTIC?

The difference between these views is partly tied to the degree to which the definition of AT is ideological. In the ideological view AT is an antidote to past trends in Western technology, particularly those of the last twenty-five years in which governments became increasingly involved in the development and promotion of new technologies such as nuclear power and space satellites, indirectly stimulating and assisting industrial innovation.

The alternate view stems from a more pragmatic definition of AT and leads to the conclusion that the whole AT movement is simply a manifestation of an increasing tendency towards diversity and pluralism in today's world. Thus, it is argued, AT will occupy an increasing number of ecological niches in the global technological system but only in places where it is adapted to its environment. Indeed the very concept of "appropriate" connotes nothing more than adaptation, and in this sense all technology must ultimately be appropriate. But the special characteristics of smallness of scale and susceptibility to community control are less important than the overall measure of adaptation to the social and natural environment, which may imply large scale and centralized control in some instances, small scale and decentralized control in others, or some symbiotic combination of the two.

We know that in the biological world ecosystems generally evolve toward diversity and more elaborate differentiation, involving a multiplicity of ecological "niches." The same has been true in the economic world so far as labor functions are concerned, but until now at least the trend has also been for single technologies to drive out alternate technologies fulfilling similar functions, with differentiation restricted to matters of style and aesthetics rather than basic technological configurations. In the following sections I shall argue that for several reasons this tendency toward "technological monocultures" is likely to be reversed in the future, leading to greater diversification among technologies with perhaps less fragmentation in the functions of the human component of the system. In other words I am projecting a shift toward greater diversification among technologies combined with less fragmentation in the division of labor in the human component.

Reasons for Technological Convergence
Rather Than Diversification in the Past

One of the dominant factors in the last century of technological development has been that of economies of scale. The more favorable the competitive economics of a given technology, the more it grows in scale of application, and the greater the scale, the lower the production or delivery cost and hence the greater the competitive advantage. Thus in the great majority of cases, competition among alternative technologies has tended to be unstable, driving the system toward complete takeover by a single alternative, whose competitive advantage continually increased with its market penetration. Moreover, this process of takeover by scale has frequently been assisted by government. In the early stages of development of a new technology—such as railroads, automobiles, airplanes, or nuclear power—the government has intervened to foster and protect the infant technology, using the argument that it offered a public benefit that merited collective support. Thus government subsidized railroads, highways, air routes, and reactor development in their early stages. Through the depletion allowance and other tax benefits, it also subsidized the development of energy resources, especially petroleum and uranium. Frequently the government subsidy has continued after the technology became mature and the infant industry justification disappeared. Such a policy further reinforced the comparative advantage of the new technology even when economies of scale were by themselves sufficient to assure its relative growth in competition with more established alternatives. This happened notably with automobiles and highways, with airlines and airports, with petroleum and natural gas and their distribution systems, and with nuclear power relative to other means of electric power generation. The actual degree of favored treatment of newer technologies and the amount of subsidy are matters of considerable debate, but there can be little doubt that the effect was present in the early stages. For example, it is probable that nuclear power no longer enjoys a social subsidy and is, if anything, discriminated against by more stringent regulations than are applied to other technologies such as coal and oil-generated electric power, but this has become true only very recently.

Even within a given domain of technology the preferential emergence of a single version has been evident. Thus in the early days of the automobile there was experimentation with a wide diversity of technologies—steam automobiles and electric automobiles, for example—but the internal combustion engine or Otto cycle eventually came to dominate the field and the whole design became increasingly standardized, with variation focused on style rather than technology. Similarly in the case of nuclear power the experimentation and diversity of the later 1950s gave way before the dominant light water reactor (LWR), to the point where the barriers to market entry of competitive reactor designs such as heavy water reactors, gas-cooled graphite reactors, or the sodium-cooled graphite reactor, became insurmountable. This happened in part because the operating experience with light water reactor technology was gained through the U.S. Navy submarine program. The economies of scale and the progress along the learning curve resulting from the head start of the LWR precluded serious competition, even though many would argue that there is no intrinsic technological or economic advantage in the light water design. Undoubtedly the dominance of LWR was also enhanced by considerable political support from the U.S. government, especially in the international market, but this may itself have been partly the result of the head start of the LWR technology.[3]

In the case of modern intensive agriculture, simplified ecosystems with a high density of plants of common genetic origin supported by a high level of external inputs—such as chemical fertilizers, pesticides, herbicides, and mechanical cultivation and harvesting—have replaced natural ecosystems involving a diversity of animal and plant species which occupy a multiplicity of ecological niches in the same geographical area. The simplified ecosystem is referred to as a monoculture, and in analogy I have coined the term "technological monoculture" to emphasize the analogy between a single dominant crop species in an artificial ecosystem and the dominance of a single technology, which, through economies of scale and learning, excludes competitive technologies whose only disadvantage may be that they had a later start.

In what follows I will bring out reasons why I think the era of technological monocultures may be on the wane. Indeed, even in agriculture, there is growing interest in the interstitial cultivation of more than one crop species and a trend toward greater ecological complexity in cropping systems and a less artificial environment.[4]

Factors Making for Technological Diversification

The circumstances that favored the emergence of technological monocultures in the past may now be changing. One of the most important changes is an apparent saturation in the benefits of scale for many technologies, and the emergence of several factors that constitute diseconomies of scale. Even in the case of the "learning curve" we find an increasing number of instances where learning turns up adverse secondary consequences of a technology as well as making it cheaper and more reliable. Thus, at the same time that learning enables us to make something cheaper, changes in design necessitated by what we learn of adverse effects may make it more expensive. I will come back to this point.

Environmental Impacts

Increases in the scale of a technology can be of two types. In one case the individual embodiment of the technology becomes larger and larger, and economies of scale result from this size. Well-known examples include electric power plants, supertankers, wide-body jet aircraft, trailer trucks, and long freight trains. Scale can also mean simply the extent of diffusion of units embodying the technology rather than the size of individual units. Examples are the population of automobiles, trucks, aircraft, television sets, and household appliances, or the penetration of the primary fuel market by petroleum and natural gas. Both of these kinds of scale exhibit diseconomies as well as economies. For example, larger power plants concentrate pollutant emissions and discharges of waste heat in a smaller geographical area, so that controlling them requires more expensive technology, such as cooling towers, pollution abatement equipment, or location remote from load centers leading to more expensive transmission facilities.[5] The effects may be not only harder to control, but also more publicly visible than would be true for a more dispersed technology, even though the total pollutant emissions might be greater in the dispersed case.

Accidents to supertankers can cause much more serious pollution incidents than accidents to smaller tankers even though small tankers and other merchant shipping are responsible for a much greater total volume of discharge than supertankers; accidents to large aircraft can cost many more lives with much more adverse public reaction than smaller accidents. Large nuclear reactors require much more safety equipment—e.g., emergency core cooling systems—and the small possibility of large radioactive releases may loom much larger in public perception than the relatively smaller and more probable accidents that contribute more to the mean or "actuarial" risk. Indeed we are witnessing today a situation in which the location of large-scale facilities of all kinds is encountering increasing political resistance. There are fewer suitable locations, and each location decision requires a longer and more contentious decisionmaking process. This political resistance can be viewed as a form of diseconomy of scale. Whether or not the diseconomies offset the economies varies from case to case. In the case of power plants, it appears that they may have, especially if one includes the political difficulties of location and the delays in construction as forms of diseconomy. In the case of large aircraft, the increase in safety and reliability has so far prevented the diseconomies from generating public resistance to their use.[6]

In the case of diffusion of a technology the diseconomies of scale can manifest themselves in environmental problems, in traffic congestion, or in more subtle social costs. For the automobile the smog phenomenon resulting from the photochemical transformation of pollutants is a nonlinear effect. Traffic congestion is highly nonlinear; when the traffic density reaches a critical value, the mobility benefit of the automobile nearly disappears. If we attempt to relieve traffic congestion with more highways and related facilities, adverse public reaction rises nonlinearly primarily because of the claims on limited land. Also public perceptions of disbenefits often seem to cross a threshold when the population reaches a critical value. This is notoriously true in the case of aircraft noise in the vicinity of airports, where public protest seems to appear abruptly when the product of noise level times population affected reaches a certain threshold value.[7]

It seems almost certain that all these resistances to further increases in scale will lead to a greater diversification and dispersion of technology in the future. Large central power plants will not disappear or stop being constructed, but there are likely to be more smaller power plants, probably using low-Btu on-site coal gasifiers in conjunction with gas turbines or combined gas and steam turbines, more cogenerated electricity in factories,[8] more integrated utility systems in conjunction with commercial developments, eventually solar energy installations, or fuel cell generators serving the same areas that electric substations do now.[9]

In transportation also there is likely to be greater diversification of technologies, especially quasipublic vehicles such as vans and other transportation systems that go under the generic name of paratransit.[10] This will not be revolution, but evolution. The automobile is here to stay, but its complete monopoly of transportation will decline, and a greater variety of vehicle types adapted to particular conditions such as downtown mobility will appear. This will happen because it will tend to relieve the diseconomies associated with technological monocultures. The secondary consequences of different technologies differ and thus have a smaller impact for a given level of primary benefit than a single technology.

Inefficiency of Organizations

The success of modern industrial society has been based on the ever finer division of labor in ever larger and more complex organizations. Here again, however, there are signs that we may be reaching a point of diminishing returns in the effectiveness of large organizations. I am not suggesting a return to the small entrepreneur and craftsman, but merely the slowing of what has hitherto been the dominant trend in the evolutionary growth of organizations, both governmental and corporate. Beyond a certain size, and despite remarkable efforts at adaptation, large organizations may cease to be sufficiently sensitive either to their external environments or to the human needs of their employees and managers. This is, in fact, just another form of a diseconomy of scale.

Many contemporary social and political trends seem to be symptomatic of these diseconomies of scale in organizations. The revolt against government bureaucracy and against big government in general is one example; the distrust of large private corporations is another.[11] But an even more important development may be the self-examination of organizations themselves, particularly as manifested in experiments on work organization and the quality of working life. These experiments aim to move away from hierarchy and status in organizations, to increase job satisfaction and heighten trust and involvement in the work. Many of the experiments to improve the human conditions of work have turned out unexpectedly to increase productivity as well.[12]

Realization of the diminishing returns associated with large hierarchical organizations is also leading to reexamination of the relation between the design of technologies and the nature of the work organization that is compatible with them. This has led in turn to the notion of sociotechnical design, the matching of technology to the human requirements of work and to more democratic and participatory work organizations.[13] It is clear that some technologies lend themselves better to democratic and self-managing work organizations than others. We are only at the beginning of these developments, and so it is difficult to see clearly what they may actually mean for technology, but it is probable that they will push against further increases in the scale and complexity of organizations and against the further division of labor. In fact the appropriate technology movement is in part predicated on disillusionment with large organizations, although in my view this represents an overreaction.

APPROPRIATE TECHNOLOGY IN A MARKET FRAMEWORK

There are two rather different ways of looking at appropriate technology. One perspective sees it as the technology that would be selected in a perfect market and maintains that technological development has become distorted in today's world because of market imperfections and intervention by government in the generation and subsidization of technology. The other views it as the technology that supports social values and goals other than economic efficiency and that is therefore probably incompatible with a free market economy. Some advocates of appropriate technology seem to argue on both sides at once; that is, they say that, rightly looked at, appropriate technology is more economic than present technologies, but at the same time they seem to argue that the inevitable triumph of appropriate technology in the marketplace has to be helped along with a large dose of social coercion of consumers who have not yet seen the light.

The market failure school of AT derives its arguments from ideas such as those in Galbraith's *New Industrial State*.[14] Here the notion is that in large corporations and government bureaucracies technology is created for its own sake to satisfy the inner drives of the organization and the technologists, and then it is forced on consumers by high-powered advertising and political lobbying. People are made to want what the "technostructure" produces, rather than the corporations producing what consumers desire of their own free will and governments responding to free public opinion. If there were real consumer sovereignty, and if there were really many relatively small producers that were competitive, the result would be something like appropriate technology. At least so the proponents of AT argue.

This argument runs up against one difficulty, namely the existence of externalities, which are not normally incorporated in the price of goods and services in a competitive market unless a way can be found to assign a market value to them through the political process or by ingenious extensions of the market such as marketable pollution permits or competitive compensation of communities for disamenities. The proponents of AT would probably argue that since in their world technology is small scale, externalities are less important, and since their effect is largely local, they are more readily incorporated in decentralized market mechanisms.

The fact remains, however, that in a free market what is being maximized is aggregate national income. Externalities can in principle be taken into account by compensation of those adversely affected while economic efficiency is retained as the criterion of performance. Environmental externalities, such as those related to health and accidents, can thus be incorporated in the system. However, when one begins to think about other social consequences such as effects on income distribution, on the quality of working life, on settlement and industrial location patterns, on such intangibles as a sense of self-worth, a feeling of community and mutual support, democracy and equality, and so on, it becomes more and more difficult to imagine how such values can be incorporated into a system of market prices.

Thus it is that the second school of AT tends to identify itself more with a socialist or planning approach than a market approach. The criteria for choice of technologies are primarily noneconomic and are supposed to support certain defined social values. These values are associated with equality (lack of hierarchy), self-sufficiency, minimum ecological impact, respect for nature, and maximum control by individuals over external factors that affect their own lives. There are, of course, internal contradictions in this approach; individual autonomy is to some extent incompatible with collective planning unless there is nearly unanimous consensus on the goals to be achieved. Such consensus seems incompatible with democratic pluralism and has really only been achieved in wartime conditions. However, the argument of the proponents of AT would be that the choice of small-scale, decentralized technologies which minimize interdependence could reduce, if not eliminate, these conflicts of value.

In many ways the discussion of AT in connection with developing countries is more straightforward than in connection with developed countries, since the LDCs are seen as having a choice. They do not have an existing technostructure to replace. Furthermore, there are obvious examples in which the application of Western technology to LDCs is uneconomic or inefficient by ordinary market criteria simply because of the gross differences in relative factor costs that exist between developed and developing countries. One can make a case that much technology used in LDCs is "inappropriate" simply because of market failures. These market failures occur partly through the intervention of governments and partly because of the dominance of the international economy by the economic institutions of the developed world. The influences of technological monoculture are especially strong in relation to the developing world, which offer too small a market to warrant the development of special products to fit it. There may thus be a strong affinity between the market view of AT and the move toward greater self-sufficiency and autonomy in the LDC economies. The less interdependent these economies are with world markets, the more likely are their choices of technology to be governed by local factor costs and considerations of local economic efficiency.

THE SOCIAL CRITERIA FOR APPROPRIATE TECHNOLOGY

In what follows I shall list the criteria usually considered as defining the social character of AT and then discuss each of these criteria in relation to various examples of technology. In this discussion, the orientation will be toward the industrialized countries and their future development rather than toward the LDCs. The characteristics usually attributed to AT are as follows:

1. It is small scale, minimizing requirements for hierarchical management or bureaucratic organizations to operate it. It is supposed to be capable of being operated by democratic or collegial organizations with minimal dependence on specialized knowledge or division of labor. In other words AT is technology requiring neither the authority of hierarchy nor the authority of expertise for its operation.

2. AT reduces interdependence generally, and particularly it reduces the dependence of individuals on the performance and intentions of people they do not know personally and cannot influence face

to face. It is thus designed to achieve a situation of maximum mutual accountability; people are compelled to be accountable directly and personally to all those who are affected by their actions.

3. AT is designed to be compatible with the preservation of nature; it has minimal ecological impact. It replaces an ethic of the conquest of nature by an ethic of living with nature.

4. AT is designed to use renewable rather than nonrenewable resources. This is often expressed by advocates of AT by saying that it uses ecological income such as solar energy rather than ecological capital such as fossil fuels, accumulated by nature over geological time periods.

5. AT is designed for decentralized management. This is really a different way of expressing criteria (1) and (2), since decentralization presumably reduces hierarchical authority as well as interdependence. Similarly AT depends on the exploitation of local resources and emphasizes local self-sufficiency, thus reducing dependence on resources controlled by people one does not know personally.

6. AT tends to be more labor-intensive than conventional technology and is designed to provide meaningful work for the whole population. Because it can be operated by people with interchangeable skills (item (1)) it does not displace unskilled labor; it generates employment.

In order to achieve goal (6) as well as goals (3) and (4) dependence on AT seems to imply the curtailment of economic growth as conventionally defined. This in effect requires a consciously enforced policy of declining labor productivity, which is to be achieved not only through more labor-intensive production of goods and services, but also through greater emphasis on personal services and cultural creations.

Let us examine each of the characteristics ascribed to AT in somewhat more detail.

1. Small-scale, Nonbureaucratic

The relation between the scale of technology and its bureaucratic management is not as self-evident as the proponents of AT would have us believe. Many technologies when considered as a total technological system are mixed in their social characteristics. Furthermore, the connection between the character of the technology and the nature of the organizations needed to manage and control it may be less close than is assumed.

If one looks back at the famous Model-T Ford, it appears to exhibit many of the characteristics ascribed to AT. It was much more completely under the control of the individual than other contemporaneous modes of transport such as the railroads. It was much more nonpolluting than its predecessor, the horse and buggy. It did not consume large amounts of resources (to feed the horses). It was more energy-efficient than the coal-fired steam engines that pulled the railroads, and it did not belch a pall of black smoke and cinders. But above all it could be repaired by any farmer's boy with bailing wire and string. It provided great freedom and personal mobility in a still largely rural and isolated society. Yet all these nice AT characteristics depended upon its being very cheap and available to a mass market, and this in turn depended on its being produced in giant vertically integrated factories on assembly lines in which hierarchical authority and fragmentation of work were carried to extremes. Thus we had a technology that at its interface with the individual consumer seemed to have all the benign characteristics ascribed to AT, but in its production and distribution exhibited in the extreme the worst characteristics of modern centralized industrial technology in a highly authoritarian social structure. These widely differing social characteristics within one technological system were complementary. They depended on each other. The integrated factory and sales system depended on a mass market, which in those days required that the car in the hands of the consumer be simple, cheap, and repairable. But its low cost depended on the technology of mass production, which required a highly structured and differentiated organization.

This paradoxical example may be important because it could be a paradigm for many of the small-scale decentralized technologies being advocated today as in the case of energy generation. Decentralized solar energy systems will have to be mass produced and distributed on a large scale if they are to be cheap enough to replace present energy sources. Much of the debate about the rate of market penetration of solar energy systems revolves around the question of how soon the "learning curve" will bring the cost down to where it is competitive. Furthermore, to achieve a mass market, solar technologies will have to be standardized and will require an elaborate distribution and service network. Given current dissatisfaction with the distribution and particularly the service associated with the automobile technological system, what reason is there to believe this will all disappear with individual household solar energy systems if they are used on the scale necessary to replace present energy sources? It would appear that as compared with conventional energy systems the locus of bureaucracy and hierarchy is merely shifted; it does not disappear from the scene.

Indeed one characteristic of modern electrical generating and distribution systems is that from the standpoint of the consumer they are practically "transparent." The householder who throws the switch does not know whether the electricity was generated from coal, uranium, wood, or water. In fact the only real interface with the generator comes via the monthly utility bill. It is this which is the focus of most of the resentment and the rhetoric about "centralized" systems; it is only here that the consumer comes in contact with the "gigantic bureaucracy" associated with "centralized" electric generating technology. However, it is not obvious that there would not be a similar bureaucracy associated with the servicing of household solar or other decentralized systems.

As for reliability, the most vulnerable parts of the present electrical system are at the periphery, not the center, of the network, despite a few isolated incidents like the New York blackouts. It is true, however, that individual systems will not all go out at the same time, except possibly in storms or earthquakes. In fact the question of vulnerability appears to be a very complex one, without an obvious advantage for either centralized or decentralized systems. For electrical networks centralization and integration increase reliability in respect to disturbances up to a certain magnitude; beyond that magnitude the rare failures that do occur may be more disruptive than any combination likely to occur with more decentralized systems. On the other hand, the vulnerabilities of the large systems can be designed against. It is only a question of how high a cost society is willing to pay to avoid or mitigate contingencies of very low probability. The inclusion of some decentralized local capacity, for example, fuel cell or solar substations, in networks served mainly by central generators might eventually be a way of combining the advantages of centralization and decentralization in a single system.

It has been observed that the evolution of information technologies—computers and communications—is such as to favor decentralization. The large central computer with satellite terminals has been replaced by the self-standing minicomputer with network connections. The typical computer-communications network consists of minicomputers at the nodes of an interconnected system. Since the processing of information is more and more at the heart of organizations, the new information technology makes possible more decentralized organizations. The subunits of the organization are not autonomous, however, but interdependent through the network. Thus the precise organization form that matches the technology may not be fully determined.[15] It could be centralized or decentralized, but social preferences will tend to push it in the latter direction. The point is that information networks are compatible with decentralized forms of organization, in which the groups at the nodes have considerable autonomy. The mode of organization has the possibility to become more horizontal than hierarchical.

Information technology also permits more complex control of manufacturing processes and even services. Standardization becomes less important; computerization makes possible the programming of serial production to vary the product along the line without shutting it down for retooling. In other words it makes possible, though it does not guarantee, that the product or service may be more closely tailored to the customer's needs or wants in the traditional manner of crafts. We are a long way from this situation today, although its possibility can be foreseen as implicit in the capabilities provided by future developments in information technology. But it is by no means determined by this technology.

Throughout the industrialized world there is increased interest in experimentation with more democratic, less hierarchical forms of work organization. Some of these experiments represent reorganization of the division of labor more or less within the present technological structure of manufacturing or service delivery, but an increasing number of experiments involve the redesign of production technologies to better meet the needs for worker satisfaction and human development in work. The best-known experiment in this regard is the Volvo assembly plant at Kjalmar, Sweden,[16] where the assembly line has been replaced by a more flexible production technology permitting greater control by workers over their own working patterns and a less finely differentiated division of labor. The work organization is less hierarchical, more of a team or collegial effort, and this is somewhat related to the characteristics of the manufacturing technology, but it is by no means a return to craftsmanship or small scale. Preliminary evaluation of such experiments indicates that an unanticipated by-product is frequently rather spectacular increases in productivity, accompanied by improvements in quality control of the product. Long-term success, however, requires that the economic benefits of productivity and quality-control gains be shared equitably between workers and management. The principle of these experiments is that independent decisionmaking power should be delegated as far down in the organization as possible. In this regard it is interesting to note that what has long been known as a good principle for the organization of research and development is apparently applicable to many other forms of organized work. Coordination in accordance with agreed upon general goals substitutes for hierarchical management with all wisdom descending from the top.

2. Minimizes Interdependence

One of the main principles advocated for AT is the reversal of the long-term trend towards greater interdependence among human groups. This principle is used as a criterion at all levels from the community to the nation. What is advocated is a dependence on local resources to the greatest extent possible, and what is condemned is the vulnerability of people to decisions made by individuals they do not know and cannot influence face to face, whether these decisions are made by the leaders of other governments, the managers of multinational corporations, the bureaucracies of the federal government, or simply through the impersonal workings of a world market. The more extensive interdependence becomes, it is argued, the more abstract and impersonal become the relationships among people and groups. More and more decisions are made by people who never see their detailed consequences for other people; more and more of the conditions of life are affected by decisions those impacted cannot control or even perceive or understand.

It is here that I depart most strongly from the advocates of AT. Interdependence is a fact of modern life, and I see little prospect of reversing it without great impoverishment of life both in a material and in a spiritual sense. In fact most of the specific appropriate technologies advocated merely substitute one form of dependence for another. However, it is also true that a mixture of different kinds of interdependencies may have advantages in reducing the *sense* of dependence. For example, having a solar energy installation or a CB radio may increase the psychological feeling of autonomy even when it does not actually substitute for the telephone or the power-line. This enhancement of psychological autonomy may be particularly important in cases where there is a large asymmetry in power between the dependent people and the organizations and people on which they depend. The point is that a large decrease in the sense of dependence can probably be purchased with a rather small decrease in *real* dependence, and this is a possibility that ought to be given more weight in the planning of technological networks.

In practice, however, the feeling of autonomy or self-sufficiency, like the feeling of privacy, is more likely to be something purchased as a luxury by the affluent than something available to the poor. Thus, to a degree, the reduction of interdependence will be experienced as undemocratic and elitist for the great majority of people. Just as it was the affluent who first owned automobiles, television sets, and pocket computers, so it will be the affluent who will first own solar heating and cooling systems or household photovoltaic generators. Whether in fact the affluent should be encouraged in these preferences through tax credits or other forms of redistribution of tax revenues is a question that has not been fully faced up to in most discussions of public policy measures to encourage the adoption of solar energy.

In the rural areas of LDCs, which lack electrification or distribution systems for commercial fuels, technologies that would be uneconomical in the rural areas of developed countries may actually be cost-effective. This is partly a consequence of the fact that the distribution systems in the developed countries were put in place at a time when they were much cheaper than their replacement costs would be today. It has been argued, for example, that there may be a market for photovoltaic generating systems in LDCs long before they are economically justified in industrialized countries with electric distribution networks already in place. There may be a conflict, however, with the ideal of self-sufficiency. Photovoltaics, to be economically attractive, require sophisticated production processes and exotic materials. Not only are they capital-intensive in terms of the amount of investment required for a given output of electricity, but they will also require substantial amounts of foreign exchange.[17]

Other ways of using renewable resources may also be capital-intensive, but if the capital is such that it can be created by local labor, thus using an abundant indigenous resource, it may still be advantageous to use a capital-intensive technology provided the other resources used are also of local origin.

Most of the population of the industrialized world lives in urban areas, and almost every projection indicates that there will be a rapid shift of population from rural to urban in the LDCs. The trend to urbanization may be slowed but is unlikely to be reversed. Furthermore, the experience of China seems to suggest that urbanization can be retarded only by extreme forms of political control over the movement of people, something that may be unlikely to continue for long even in China. The advanced countries are already more than 80 percent urban, and although this trend has now slowed, and even reversed in a minor way, the settlements in nonurban areas and small towns are highly interdependent with the metropolitan centers. Cities are inevitably characterized by interdependence, and as between the DCs and LDCs cities are much more alike in their technological character than rural areas. For example, the average per capita consumption of commercial energy in New Delhi is about one-half that of a typical Western European city like Dusseldorf, while the energy consumption on a per unit area basis is considerably larger than in Western cities because of the higher settlement density.[18] Such population agglomerations can probably be served only by largely centralized technologies and associated distribution networks. Similarly transportation needs and other municipal services such as water supply and waste disposal are best served by public transit and by other integrated and centrally managed facilities. But all such services to urban dwellers inevitably put the citizens more at the mercy of decisions made by public employees, as transit strikes in the city of New York have amply illustrated. Yet, paradoxically, many of the advocates of AT cite the automobile as an example of an inappropriate technology and see public transit as the technological wave of the future. While it is possible to foresee the growth of cities leveling off, or even declining, the trend toward urbanization throughout the world seems unlikely to be reversed drastically, and this implies at least the present degree of interdependence, with at most some greater self-sufficiency at the margins. But these more self-sufficient services are likely to be accessible only to the more affluent or privileged segments of the population, as suggested previously.

3. Minimizes Ecological Impact

The reduction of the ecological impact arising from the large-scale deployment of technology is a new and important objective of technological development, but it is not clear that the other characteristics ascribed to AT are necessarily and inherently compatible with this one. It would appear, for example, that centralized technologies lend themselves to the minimization of *total* ecological impact more readily than decentralized ones. In the United States, cities occupy only 3 percent of the land area; what they do occupy is largely "ruined" from an ecological viewpoint. Yet the aggregation of people into cities probably on balance reduces the ecological pressures on the rest of the land area. Thus cities, like other centralized and specialized facilities, can be thought of as devices for reducing the *net* adverse ecological impact of economic progress. As another example, nuclear power plants and facilities associated with the nuclear fuel cycle require less land than any other power source, and the ecological damage can be minimized by proper design. It is true that transmission lines also use land, but we can look forward to a time in the future when underground superconducting cables will replace present overhead lines. Indeed, the economic competitiveness of new methods of transmission depends to a considerable extent on the centralization of both energy production and load centers. Nuclear power plants require far less movement of materials in and out of the plant than is the case for other present power sources, and this will be even more true if breeder reactors replace the present generation of converter reactors.

A great deal of confusion arises because of the different scales of different technologies. A solar energy unit designed to serve a single household or a small community seems ecologically very benign, but when one considers the materials requirements for the solar power equivalent of a 1000 GWe nuclear plant, the ecological benignity is less self-evident.[19] Yet that is the proper comparison. It has been estimated that to provide the energy equivalent of a 1 GWe generating plant by cutting and burning wood would require the entire forest area of New England if the wood were to be supplied on a self-sustaining basis; yet the New England states now use more than 8 GWe of electrical power.[20] This is not to denigrate wood as a source of fuel in appropriate circumstances; much can be done, for example, to make paper plants self-sustaining in energy by proper use of wood wastes to generate electricity and process heat for the plant, thus reducing solid and liquid wastes at the same time.[21] It has been estimated that agricultural and municipal wastes could be processed and used to produce electric power that would be equivalent to 5-10 percent of current electric energy consumption.[22] This should certainly be done, but to do so successfully is likely to require centralized processing and a highly organized collection system. It is much easier, for example, to derive energy from the animal waste in cattle feedlots than to collect it from dispersed farms.

Not all ecological benefit derives from centralization, however. Cogeneration of electricity and process heat for industry can result in more efficient use of primary energy resources, and hence in less ecological damage, than central generation far from a place where the waste heat can be used. However, it may be more difficult to control emissions of many small cogeneration installations. The point is that there is no necessary correlation between either centralization or decentralization and ecological benefit. This can come out either way, depending on the particular technology. For example, one ecological benefit might come from the substitution of sophisticated communications for transportation, especially individualized transportation such as the automobile, yet this will probably require a high degree of centralization and bureaucratic management. Ecological benefit is something that must be pursued in its own terms as one of the criteria for technological choice, which in practice will not be much correlated with the other desirable criteria usually attributed to AT.

4. Promoting the Use of Renewable Resources

Although the energy resource used in solar power installations is "free" and indefinitely sustainable, the other resources required to capture and transform the sun's energy are for the most part nonrenewable and may indeed be required in larger amounts than is the case for other energy technologies depending on nonrenewable fuels. Today, in fact, it appears that there is a trade-off between the nonrenewability of the energy resource and the amount of other nonrenewable resources required to utilize it. Natural gas appears to require less investment to produce, distribute, and burn than any other energy resource, yet at present it appears to be one of the most rapidly depleting, at least in the United States. Solar power seems to require the greatest use of other nonenergy minerals, although this cannot be said with complete confidence in the absence of more detailed engineering than has been carried out to date.[23] Were biomass to be utilized on a large scale through "energy farms," nonrenewable resources in the form of fertilizers, harvesting machinery, and, to some extent, land would have to be employed on an unprecedented scale. Most studies indicate that, beyond the use of marginal and otherwise unproductive land or organic wastes, biomass as a source of energy is ecologically costly, perhaps exceeded only by hydroelectric power in its prodigal use of irreplaceable resources.

The foregoing strictures on the complementary use of nonrenewable resources in connection with the use of renewable energy resources apply only when one is talking about the complete or nearly complete replacement of existing energy sources by renewable ones. There exist many technicoeconomic "niches" (analogous to ecological niches) in which renewable energy resources can make a useful contribution. A typical example is the local production of natural gas from agricultural wastes, seaweed, or otherwise useless plants in rural areas. Solar energy, wind, small-scale hydro, and wood could serve rural and remote areas effectively under certain conditions, but the balance of ecological impacts would have to be carefully evaluated. These developments should be encouraged where they make economic and ecological sense but not under the illusion that they represent a complete alternative to existing energy sources. If such technologies are pushed beyond the areas and situations in which they are demonstrably advantageous, they are likely to be both economically costly and ecologically damaging, much as the use of firewood in many LDCs bids fair to become a global ecological disaster contributing to desertification and soil erosion. In effect what I am saying is that AT must be truly appropriate to the circumstances in which it is used, but that its appropriateness in comparison with more traditional technologies is likely to be quite limited geographically, socially, and ecologically.

5. Decentralized

I have already taken up the criterion of decentralization, which is really a combination of the criteria of small scale and minimal interdependence. The point was made that modern technologies were more versatile with respect to centralization and decentralization than the technologies of the past, but that decentralization was probably not a goal worth pursuing for its own sake. Indeed decentralization is almost impossible to define when dealing with a concrete technological alternative such as an energy, transportation, or water system.

6. Labor-Intensive

The emphasis on AT as labor-intensive has derived mainly from the perception of the growing amount of surplus labor in the LDCs.[24] The one respect in which the communist societies seem to have performed better than the capitalist polities is in putting this surplus labor to work. This has been notably true in China in contrast to, say, India. However, as unemployment has become a growing problem in Europe and the United States, there has been a tendency to recommend labor-intensive technologies as a prescription for the developed countries as well. In the long run this would be consistent with curtailing economic growth as a means for minimizing ecological impact. If growth is to cease without massive unemployment, the historical growth in labor productivity and labor force participation will also have to cease. This could come about through decline in the rate of technological change, through a decline in savings and hence of investment, through a shortening of the work week, or through a shift to personal services accompanied by lower per capita goods consumption, or by some combination of all of these. It has frequently been argued that employment and energy consumption may be substitutable, since the production of energy is capital-intensive and currently absorbs about 20 percent of capital investment in the United States.[25] If the same investment were shifted to less capital-intensive technologies while technological innovation were concentrated on improving energy efficiency—getting more output for less energy—the economy could reach full employment, or at least so it is argued.

The situation is complicated and rather unclear. All production of energy is capital-intensive, but the construction process for energy-supply facilities is about average in capital intensity, so that as long as energy consumption is growing, the effect on employment overall is uncertain. For the most part econometric models suggest that achieving a given level of GNP with less energy consumption has a slightly positive influence on total employment, but this conclusion is based on many assumptions about the economy that can be questioned by skeptics.[26] Saving labor in one technology may release labor for employment in more productive areas; hence one cannot draw conclusions about employment effects from looking at particular technologies.

Similarly, proponents of continued economic growth would argue that the question of ecological damage should be faced in its own terms, not by curtailing growth unselectively. Environmental quality can be viewed as a good which becomes incorporated in the measures of GNP through internalization of the costs of pollution abatement.[27] Indeed, the pollution abatement and environmental monitoring industries have been a major source of new employment in the United States during the 1970s. It is possible that the costs of improving environmental quality may retard the growth rate, but probably the principal effect is to shift the composition of the GNP toward the production, operation, and enforcement of environmental controls. To the extent that application of controls is implemented by government, this also implies some shift of GNP from the private to the public sector. Also to the degree that the public sector productivity growth is evaluated as zero in national income accounting, the average productivity growth, and hence GNP growth, is reduced. Unless slower economic growth is regarded as intrinsically desirable, deliberate selection of technologies that do not optimize the use of factors of production between labor and capital simply reduces overall welfare. In a well-functioning economy labor saved in one area can be applied to other areas, although lack of labor mobility and related structural problems may limit the rate at which this can happen in practice.

The problem with using labor-intensity alone as a criterion for the selection of technology is that any reasonable criterion should include the *quality* as well as the *quantity* of labor. By quality I refer not only to the requisite skills and education but also the quality of the work experience itself. Surely it is not desirable to bring back drudgery and back-breaking physical labor merely for the sake of generating more employment! However, if both quantity and quality are to be used as criteria then a problem of trade-off between the two arises. In many cases, for example, especially with modern information technology, automation tends to improve the quality of labor while reducing its quantity. Supervision of an automated chemical process in the chemical industry is surely a higher quality job in both of the aforementioned senses than the batch mixing of chemicals in a hand-crafted chemical operation. Operating a back hoe is surely a more interesting and rewarding job than digging a ditch with pick and shovel. Operating a combine is more humanly rewarding than scything a field of wheat. The advanced thinking on humanization of factory work going on in Sweden suggests that combinations of complete automation of some processes with a craft organization of others may be more humanly meaningful than either one alone. The dangerous, hot, and dirty operations should be removed from human contact completely, while such operations as assembly and inspection may benefit from a more autonomous handicraft type organization.

This again illustrates the point made at the beginning of this chapter that the proper place for AT is in "niches" that are symbiotic with more conventional technology. The choice of technology cannot be subjected to generalized rules, but must be arrived at on an evolutionary, experimental basis in which a number of criteria are combined.

There is no argument with the fact that the mix of capital-intensive and labor-intensive technology should generally be such as to make optimum use of the correctly valued factors of production in the long run. The only issue is what constraints with respect to other impacts on environment, health and safety, quality of work, and general social impact should be applied, especially in the light of a high degree of uncertainty attending the prediction of some of these impacts. This statement is important because in many cases political influences have distorted factor costs and directly influenced technology selection so that the final mix of technologies has not been economically optimal for the circumstances. It has often been pointed out, for example, that less-developed countries have a bias toward unsuitably capital-intensive technologies, largely because of their ready availability in packaged form from developed countries and the social prestige associated with the most advanced technologies. In developed countries social policies that make labor more expensive or make it difficult to discharge redundant labor result in the substitution of capital for labor, and sometimes in the movement of labor-intensive operations offshore to cheap labor areas. The second- and third-order consequences of policies regarding labor are difficult to anticipate and hence to control. Thus the encouragement of labor-intensive technologies is not likely to be successful unless other steps are taken to make labor cheaper relative to capital and other resources, but this encounters serious political resistance. The wage-earner is not very enamored of no-growth policies that affect his real disposable income directly, and yet that is what the adoption of labor-intensive technologies that do not employ all factors of production optimally actually implies.

SOME GENERAL CONCLUSIONS

Appropriate technology is best considered as a complement rather than an alternative to the traditional centralized technologies characteristic of modern industrialized societies. Deployed in certain "niches" within the existing technological structure, it may provide benefits that could not be realized with either AT or conventional technology alone. These benefits are likely to be realized soonest by the more affluent.

There is evidence of diminishing returns to scale in connection with a number of large-scale centralized technologies, and this means that smaller-scale technologies are likely to be relatively more competitive than in the past, especially for particular applications in areas of low population density and where existing distribution networks for more traditional technologies are sparse.

The desirable social characteristics usually attributed to AT are frequently not compatible with each other; for example, the proliferation of small-scale technologies that offer the consumer greater autonomy may require mass production, distribution, and service that can be carried out only with large organizations.

Many of the desirable characteristics associated with AT may be achieved by evolutionary modification of existing technologies and associated administrative structures and work organization, without necessarily making a radical modification of the basic technology.

Cheap and sophisticated information technology, characterized by rapidly declining costs per unit function, makes possible flexibility and individualized control of existing technologies. This in turn facilitates the change of working and administrative structures, which satisfies several of the goals of AT.

In the absence of a radical depopulation of the world, an increasing proportion of people will live in urban or densely populated areas in the future. Conurbations and dense settlements are inherently dependent on centrally organized technologies, distribution networks, and services. It would appear that AT has relatively little to offer to this fraction of the population in comparison with more traditional technologies.

The ecological impact of AT may be much larger than first appears if it is applied on a scale sufficient to serve a major fraction of the world's population.

The cause of developing and diffusing AT would best be served by deemphasizing its ideological aspects and working on the best possible adaptation of all technologies to their circumstances of use, irrespective of whether they fit the definition of "appropriate" or not.

NOTES TO CHAPTER THREE

1. E.F. Schumacher, *Small Is Beautiful: Economics as if People Mattered* (New York: Harper Colophon, 1973). Also Amory B. Lovins, "Energy Strategy: The Road Not Taken," *Foreign Affairs* 55 (October 1976): 65-96.

2. "Soft vs. Hard Energy Paths," *Energy Perspectives* 77, no. 3 (New York: Edison Electric Institute, 1977).

3. I.C. Bupp, "The Nuclear Safety Controversy" (Seminar paper presented at the Harvard Business School, December 10, 1975); I.C. Bupp and J.C. Derian, "Nuclear Reactor Safety: The Twilight of Probability" (Paper prepared for Management beyond Business Project, Harvard Business School, December 1975).

4. *Consultative Group on International Agricultural Research* (New York: CGIAR, 1976).

5. National Academy of Sciences, *Technology: Processes of Assessment and Choice*, Report of Committee on Science and Public Policy, National Academy of Sciences to the Committee on Science and Astronautics, U.S. House of Representatives (Washington, D.C.: U.S. Government Printing Office, July 1969).

6. B.K.O. Lundberg, *Speed and Safety in Civil Aviation*, The Aeronautical Research Institute of Sweden Report Ho. 95—The Third Daniel and Florence Guggenheim Memorial Lecture (1963).

7. National Academy of Sciences–National Academy of Engineering Environmental Studies Board, *Jamaica Bay and Kennedy Airport: A Multidisciplinary Environmental Study*, vols. 1 and 2 (Washington, D.C.: National Academy of Sciences, 1971).

8. R.M. Ansin, Chairman, *Cogeneration: Its Benefits to New England*, Final Report of the Governor's Commission on Cogeneration, Commonwealth of Massachusetts (October 1978).

9. Arnold P. Fickett, "Fuel-Cell Power Plants," *Scientific American*, December 1978, pp. 70-76.

10. National Academy of Sciences Transportation Research Board, "Paratransit Services," *Transportation Research Record* 650 (Washington, D.C.: National Academy of Sciences, 1977).

11. Michael Maccoby, "Trust Is Also the Business of Business," *New York Times*, March 31, 1978.

12. Michael Maccoby, "Changing Work, The Bolivar Project," *Cambridge Institute Working Papers*, Summer 1975, pp. 43-55.

13. P.G. Herbst, *Socio-Technical Design: Strategies in Multidisciplinary Research* (London: Tavistock, 1974).

14. John Kenneth Galbraith, *The New Industrial State* (New York: New American Library, 1967).

15. Umberto Pellegrini, *Evolution of Computer Systems: From Centralized to Distributed Structures* (Ivrea, Italy: Olivetti, 1975), pp. 123-29.

16. Swedish Employers' Confederation, Technical Department, *Job Reform in Sweden, Conclusions from 500 Shop Floor Projects*, trans. David Jenkins (1975).

17. R.S. Eckaus, *Appropriate Technologies for Developing Countries* (Washington, D.C.: National Academy of Sciences, 1977).

18. W. Hafele and W. Sassin, "The Global Energy System," *Annual Review of Energy* 2 (1977): 1-30.

19. H. Inhaber, *Risk of Energy Production*, Atomic Energy Control Board Report no. 1119 (Ottawa, Ontario: AEC8, March 1978): cf., however, J.P. Holdren, *Nuclear News* 22, no. 5 (March 1979): 25; H. Inhaber, *Nuclear News* 22, no. 4 (March 1979): 25-26; J.P. Holdren, *Nuclear News* 22, no. 5 (April 1979): 32-34.

20. M.W. Goldsmith et al., *New Energy Resources: Dreams and Promises* (Energy Research Group, March 1976), pp. 12-24.

21. E.P. Gyftopoulus, L.J. Lazarides, and T.F. Widmer, *Potential for Effective Use of Fuel in Industry, A Report to the Energy Policy Project* (Cambridge, Mass.: Ballinger, 1974).

22. National Academy of Sciences, *Domestic Potential of Solar and Other Renewable Energy Sources*, Supporting Paper No. 6 of the Committee on Nuclear and Alternative Energy Systems (CONAES) (Washington, D.C.: National Acadamy of Science-National Research Council, October 1979).

23. Inhaber et al., also S. Baran, "Solar Energy—Will It Conserve Our Non-Renewable Resources?" (Memorandum distributed by Atomic Industrial Forum, 1978).

24. J.P. Grant, "Marginal Men: The Global Unemployment Crisis," *Foreign Affairs* 50, no. 1 (October 1971): 112-24.

25. Duane Chapman, "Energy Conservation, Employment and Income," Cornell Agricultural Economics Staff Paper, No. 77—6 (Ithaca, N.Y.: Cornell University, May 1977).

26. Energy Policy Project of the Ford Foundation, *A Time to Choose: America's Energy Future* (Cambridge, Mass.: Ballinger, 1974), Appendix F. pp. 493-511.

27. Harvey Brooks, "The Technology of Zero Growth," *Daedalus* 102, no. 4 (Fall 1973): 139-52; also Marc Roberts, "On Reforming Economic Growth," *Daedalus* 102, no. 4 (Fall 1973): 119-37.

Chapter Four

Beyond Appropriate Technology

John D. Montgomery

Only a decade or so ago it was possible (even against powerful evidence to the contrary) to indulge in the wishful belief that technology could be ethically neutral (a screwdriver can be used to repair washing machines or to commit murder, but its inventor is free of praise or blame in either case). It would still be convenient to believe that discoveries can be *applied* as rationally (or as serendipitously) as they are *made*. If so, the social choices involved in applying technologies could be made professionally, in the same objective, cool, and detached fashion in which scientists are presumed to work. There would be a clear-cut division of labor, in which specialists (politicians? economists? policy scientists?) would accept the responsibility for the social impact of new knowledge, and scientists would stick with discovering and developing it.

But such beliefs are increasingly hard to sustain. If ever social values could have been separated from scientific fact—if a Leonardo could once have designed tanks and submarines without worrying about their production and deployment—few believe they can be now, in an age of nuclear weapons, or even of atomic power, food preservatives, insecticides, power steering, high-yielding varieties, supersonic transports, or hair bleaches, each of which has social debits as well as benefits. The problems and the choices are becoming urgent in fields where once the values of progress seemed obvious: popular demonstrations at airport and nuclear plant sites in developed countries do not exempt the scientists from the equations of blame associated with technological breakthroughs.

In less-developed countries (LDCs), scientific choices are especially significant because the options are so severely limited by resource constraints. In addition, the external pressures are especially great and the time horizons have to be distant because so many decisions made today are irreversible in the foreseeable future. The need for analyzing the social impact and conflicting values of technological change in these contexts far outruns the capacities for such analysis. Thus the ethical challenges their own activity poses to scientists are both frequent and serious. But neither their own preferences and traditions nor the political environment in which they work helps scientists meet that responsibility.

VALUES

As an intellectual discipline, technological assessment, or the "choice of technique," is showing some signs of progress. The early insights about the "technologic misfit"[1] and the need for social and technological "congruence"[2] have given way to more systematic analysis. The issue of "factor proportions" has been applied critically and rigorously to industrial technologies in the hope of achieving an appropriate balance between capital-intensive and labor-intensive methods.[3] Attempts to add other dimensions beyond those of capital- and labor-intensiveness have followed,[4] and the variety of social considerations introduced into the models of technology assessment have made it the equivalent of a social movement.[5] Each step of the journey has increased the range, if not the power, of the analysis. It would no doubt be possible today to attach weights to each human value affected by a proposed technological innovation and to emerge with a calculus of the potential social costs and benefits of any choice, as seen by the groups most likely to be concerned with it. But the movement has been arrested almost before it has begun. Unanswered questions are threatening to leave technology assessment a mere intellectual pastime. What should be the denominators of value? How do the group interests add up? To such questions technology assessment has no answer.[6] And its findings are consequently having little independent effect on ultimate decisions about technological innovation.

There is as yet no convincing basis for fixing the priority among human values; even the taxonomies of value differ. The hope of performing systematic value analysis persists, however, as further investigation produces equivalencies among them. The categories of choice are repeated often enough in the major ethical formulations, and they offer enough similarities among the leading theories, to yield plausible comparisons among alternative courses of social action. Discovering anything approaching convergence in the values affected by science and technology would encourage more systematic thinking about the social choices that have to be made.

Most ethical systems assume that values lie within the range of individual preference. Individuals can create and make use of social instruments to advance their value status or to prevent their deterioration. Individuals can also reject social instruments, since theoretically they may choose to opt out of society itself if the costs of its actions to their own values outweigh the benefits. Because of such possibilities, society (in its liberal political forms) usually takes heed of individual challenges to its preferred values when they reach uncomfortable levels of intensity. Political leaders recognize, if not a social contract, at least the obligations they owe their constituents in return for continued support. There is a link between how governments use knowledge (and other resources) and their own survival.

In the twentieth century, science and technology provide some of the most important issues of public morality; few of the questions dividing mankind are unaffected by discoveries of new products, new weapons, new means of communication, new methods of production. Emigration, strikes, guerrilla warfare, and apathy have all been, on occasion, responses to misapplied technology. And they have also been the occasion for new technological solutions through the development of marginal lands, improved industrial safety, better weapons systems, and more pervasive mass communications, respectively. The trouble is that each such technological "fix" permits a few individuals to manipulate the system still further, usually to the disadvantage of the disaffected populations. Technological innovations that create disadvantages to unrepresented groups in society may benefit their sponsors and still present a serious ethical problem for scientists.

Because of such conflicts, an important role of government in all Western traditions is to achieve a viable equilibrium among social institutions that are in competition over the balance of human values in the community. The process by which this equilibrium is sought requires a continuous search for acceptable relationships among social institutions affecting values. Governments intervene when the imbalance becomes, for any reason, a political problem. But governments, perhaps especially in the less-developed countries, do not fully represent the groups most likely to be placed at a disadvantage as a result of developmental change. Effective political action is rarely available to the very social groups that have least control over the development and introduction of technology. The greater a nation's dependence on technology for achieving its collective goals, therefore, the more serious is the potential ethical challenge of scientific activity.

The first step in considering the range of effects technology may have on human values is to identify and classify the potential issues. There are many existing formulations of human values that might be used as a starting point. The United Nations Declaration of Human Rights, however, enumerates the most generally accepted list, and it can be used very readily as a basis for a preliminary screening of human values because each element of the formulation represents an arena in which citizens' claims can be advanced and appraised. Moreover, this statement of values, though somewhat eclectic in origin, can be collapsed into eight fundamental categories[7] to be used in considering the potential effects of science and technology: security, knowledge, material welfare, psychological and corporal well-being, self-expression, community and group loyalty, respect, and morality or rectitude.[8] There is no assumption of *uniformity* of individual preferences *within* each category, and the list is also neutral as to the *priorities* to be assigned *among* the categories.

In analyzing the effects of technology choices on these values, it is necessary first to acknowledge that each individual aspiration within the range of human values can be affected positively or negatively by scientific discovery and application. Both the evidence for this assumption and the issues it poses may be stated simply:

1. Man's desire for *security* (and its converse, his desire to control others) is affected by all discoveries that change the distribution of power among individuals and nations. Which "side" shall science serve—domination or protection—and under what circumstances?

2. His thirst for *knowledge* (and its converse, his desire for secrecy or exclusive possession) is the source both of scientific endeavor and of efforts to limit access to its fruits. Who should have access to knowledge?

3. His desire for *material goods* (and its converse, the impulse to surpass the possessions of others) is well served by technology, but in excess this desire also inflicts ecological injury suffered by all members of society. Which consumer goods should be supplied, and which restricted?

4. His need for *comfort* is a value that is widely sought and rarely denied to others *except* when individual and social needs cause him to inflict punishment or injury upon an enemy. Thus medi-

cal, pharmaceutical, and psychological technology contribute both to health and to torture, drug abuse, and involuntary confessions; where should science and technology draw the line?

5. His *self-expressive* and aesthetic desires are served through the exercise of skill and self-discipline. Technology can enhance these qualities through the inventions of new instruments, tools, and media, but it can also replace them through obsolescence and other atrophying innovations. What should the scientific community do when both results are observed?

6. The human values of loyalty, affection, and *community* are associated with the desire "to belong," and they are both aided and undermined by technology. The inventions associated with mass communications have both joined and isolated humanity; should radio and television ennoble as well as vulgarize? Can the disciplines of industrial society produce community as well as alienation?

7. The values of self-esteem, recognition, and *respect* are sometimes acquired at the ironic cost of denigrating others. They have been both served and challenged by discoveries of psychological and genetic differences among individuals, races, and classes. Should scientists seek ambiguity on these issues at the cost of manipulating knowledge in order to avoid the risks of bigotry and ethnic stridency?

8. The ultimate restraint value, *rectitude*, may be man's desire to be, and to seem, moral and righteous. It may also be the only value in the policy scientist's panoply that is not directly affected by some phase of science and technology activity, though its theological context has at times been severely threatened by discovery. Should scientists compromise on such discoveries until religion has a chance to catch up?

All of these values are served by social institutions, each of which in turn is likely to have preferences regarding the development and use of technology.

Recognizing the variety of human aspirations within these categories, psychologists and sociologists have sought to discover patterns among them that appear to be most compatible with scientific or technological standards. This search has enabled them to identify the characteristics of "modern man" and to use these characteristics to specify the social "pre-conditions to modernization."[9] In order to achieve such preconditions, technology is expected to introduce the disciplines and values of modern life to any society that aspires to them. But a dilemma appears as soon as these transitional features appear: the introduction of science and technology either immediately benefits certain individuals or groups whose values are most compatible with the behavior required in "modern society," or else it demands changes in the value structure of individuals and groups that desire to take advantage of the new opportunities for self-improvement. Are science and technology, therefore, to serve the privileged few whose values and resources are compatible? Or, on the other hand, are they to manipulate divergent values so that all may take advantage of the opportunities that may be created by them?

The response to this dilemma has been ambiguous. If only a few farmers have benefited from the Green Revolution, scientists have sought new technologies for those left behind. If an innovation produces a slow popular response, the tendency is to raise the level of government activity, in order to achieve the desired output without mobilizing the public's acceptance of change.[10] In short, during the course of "modernization," values do undergo change, but they do so slowly, and the core is resistant to assaults from reason or from new experience. Individuals prefer to live inconsistently, suffering "cognitive dissonance," rather than to make conscious realignments in their value preferences. Science policy must therefore either ignore or seek to accommodate the values of individuals (taken collectively) who remain holdouts against change or who are injured by it. When this problem becomes a conscious issue of science policy, the usual outcome is an attempt to create mechanisms for discovering the sources of resistance, and if the conflict appears irreconcilable, to seek new, "trade-off" options. More often, however, the response is: full speed ahead; devil take the hindmost.

The first intellectual task accepted by scientists has been to find suitable accommodations to the values of a modernizing society that are slow to change by introducing "appropriate technology." That is the minimalist approach: it identifies the characteristics of a technological change that will inflict the smallest feasible injury on most members of a society. It seeks to minimize unemployment, for example, or the outflow of capital, or damage to the inhabited environment. It can provide sophisticated answers by applying such tests as Paretian optimality to social choice in order to seek the "least injury" (as in the location of a nuclear power plant in order to reduce community protests to an absolute minimum).[11] And in cases where a choice is irreversible, surely the decision does justify at least some analytical effort on the part of scientists and technicians participating in the planning process to identify the distribution of expected costs and benefits. Moreover, this task can be rationally expanded until the analysis is as complex and diverse as our contemporary understanding of human values.

The models now available for such analysis are sophisticated enough to incorporate almost any combination of value preferences into the equation of choice. What more is needed? Is identifying the least injury to the most people sufficient? Unfortunately, it is not. If it were, defining the minimalist position on social choice would be little more than a problem of good staff work.

ELITES

The social utility even of good staff work is limited by the attentiveness and commitment of the decisionmaker to whom it is presented. Visitors to almost any national ministry in the Third World have observed and sympathized with the plight of its top professional officials: they are harassed and overburdened by unfamiliar problems that force them to engage in constant negotiations with counterparts and colleagues. Their opportunities for any serious review of long-term problems depend on the political volatility of their environment and competing demands on the slim ranks of talented technicians available to take over when they leave off. The principal cause of the frustrations so frequently encountered by staff analysts whose best work is ignored at decisionmaking time is not so much that their ideas are rejected as that they are not even considered.

One result of this official neglect is an underground network of good but unapplied staff papers. An unsuccessful program developed in one country often appears as the basis for a working paper in another country, and its most advanced version may be finally adopted in a third. International consultants and policy scientists serve as the conveyors and preservers of these untested staff papers until their ideas, approaches, and methodologies develop a life of their own. Such recognition is not as good as adoption, but it is better than oblivion. Each rehearsal of an untried proposal gains something by way of renewed assurance and a professional constituency; decisionmakers have been known to adopt on the rebound from other countries proposals that originated from their own staffs and then reached them again via the international circuit. In the post-Xerox age, policy analysts never know where their work will wind up. Thus participation by scientists in studies of the social consequences of their technologies is a casting of bread upon the waters.

The reason this underground of unsuccessful staff papers is so important in social choice is that it significantly augments the capacity of decisionmakers in the Third World to consider new options. Decisionmaking in less-developed countries usually suffers from the small and narrow bases of their elite structures, which lack both versatility and diversity. Social and political organizations in the Western democracies provide policy guidance, sanctions and restrictions, and replacements and support to decisionmaking elites, but in the LDCs they do not function so effectively as instruments of social choice because their decisions are so rarely made in a public forum. As a result, the staff work in these countries, with their widely different perspectives and traditions, provides for the decisionmaking process something of the same service of proposal-generating and coalition-building that different interest groups do in the Western democracies. There are not circulating elites in the Third World so much as there is an elite circuit for new ideas and programs. It would be instructive to trace the flow of innovative ideas about rural development from East Pakistan to Tanzania, and from Indonesia to the Philippines, or even from Maoist China to Ethiopia, by studying the proposals that were unsuccessful in the short run. They would tell as much about how social choices are made as our more celebrated studies of the diffusion of technological innovations by multinational corporations.

Elite decisionmaking and the international flow of staff work affect the role of the scientists in the formulation of technology choices. Most countries have, it is true, social institutions that embody the values previously listed (pp. 82−83), and all are potential patrons or opponents of applied science and technology. The institutions are, respectively, (1) armies and police forces to provide for collective and individual security; (2) universities to protect higher learning and the search for knowledge; (3) banks, insurance companies, and corporations to serve as custodians and mobilizers of wealth to produce consumer goods; (4) hospitals and medical professionals to restore health and well-being to the community; (5) museums, folk traditions, and trade schools to encourage and promote self-expression; (6) unions, farmers' associations, extended families, and ethnic organizations to serve the needs of workers and other citizens who desire communal association; (7) courts to protect individual rights and uphold the privileges conferred by law; and (8) churches to symbolize righteousness and remind members of their moral obligations. The full panoply of value-serving institutions is on display in every country where science itself is allowed to exist. But these institutions, often small in proportion to the population, act in isolation from each other, each pursuing its own ends and serving its own vision of social values, rarely interacting with other institutions in search of higher resolutions of values in conflict, and even less frequently becoming involved in the technological issues of social choice that have dominated American and European discussions of science policy in recent years. Scientists concerned with social choices find little intellectual guidance or organizational support from the value-serving institutions of the Third World.

Most countries, even among the least developed, have established political processes for considering social values and assessing national priorities. Yet the concept of the public interest has not emerged as a significant force affecting these processes. The succession of private interests that attempt to colonize the public sector in the LDCs has not provided an arena in which values clash so much as a battleground for different claimants to power.[12] It is rare that issues of science policy are identified at all in the political process, even when major social costs and benefits are at stake. Instead these issues are characteristically decided in response to pressures from institutions that are organized around value sectors—the same institutions that resist efforts to subject their decisions to the larger political process.

For these reasons national formulations of social values in the LDCs tend to be limited in scope and lacking in precision. The current economic plans for Korea, Sri Lanka, and Ghana, and three successive Indian statements of national purpose represent six such formulations. Taken individually, they seem conventional and even uninteresting, reflecting their origins in limited elite interaction and institutional depth. But when compared (Table 1), differences emerge that reveal important regime characteristics. These differences reflect elite judgments rather than abiding social or national purposes. It is surely significant that the three successive regime statements from India did not assign top priority to the same value. Moreover, even agreement among them can be misleading, since the same formulation has such different meanings in different contexts. The enlightenment values—knowledge generation and the development of application of science—certainly involve different uses of research and development in Gandhi's period from those sponsored by Nehru and the current Indian plan. Thus India alone among the four countries assigned priority to those values, but the implications were so different in these three periods that they could have provided little guidance to the science community in choosing among alternative technologies. Neither institutional representation of values nor the political processes of decisionmaking permit scientists to inject their views of social choice in the design and application of technology.

The task of scientists in LDCs who must consider problems of social choice is on the whole more difficult than that of their counterparts in the Western industrialized countries. They find political

Table 1.   Value Priorities in Six National Plans.[a]

	Korea	Sri Lanka	Ghana	India (Gandhi)	India (Nehru)	India (Current Plan)
National security	1	0	0	0	0	0
Economic development	2	1	(1)	(2)	1	2
National solidarity	3	3	(1)	0	0	0
Consumer goods	S	0	S	(2)	2	1
Employment generation	S	2	S	(1)	S	S
Knowledge, science	S	0	0	(1)	3	3
Individual self-respect	0	0	S	S	0	0

[a]The numbers represent formal or implied order of priority. S means a subordinate, but not ranked, value. 0 means not mentioned. ( ) represents shared place in priority listing.
Sources: Essays by Hyung Sup Choi, Robert Dodoo, and V.V. Bhatt published in this volume and an unpublished paper by D.L.O. Mendis entitled "Technology Needs in the Light of National Development Models—Sri Lanka," presented at the Wingspread Symposium on Social Values and Technology Choice in an International Context, June 1978, Racine, Wisconsin.

decisionmakers inaccessible through the channel of "normal" staff work; they can locate few active organized interest groups to which they can express their preferences and reservations. Moreover, the types of social institutions that articulate values in the Western democracies are isolated and narrowly focused in the LDCs; there is no tradition of public interest to which scientists can relate their concerns; and official formulations of national goals and purposes are transitory and often inapplicable to the issues and dilemmas posed by scientific advance. If their perceptions of the costs and benefits of a new technology are to come to the attention of public and private officials, scientists will have to adopt roles and take upon themselves responsibilities that are heavier and more active than those of their counterparts in Europe and America.

ROLES

Although most LDCs are elite-dominated to a much greater degree than are Western industrialized societies, government decisions are not entirely determined by elite self-interests. It is true that coup leaders tend to leave office rich if they do so at all; that many government-distributed benefits go directly or indirectly to their allies and to counterelites who have to be placated; and that the rural and urban poor in most of Asia, Africa, and Latin America are underrepresented in the councils of government and cannot mobilize the efforts of scientists and technicians who are discovering or applying new knowledge in their community. But in spite of these disturbing patterns of policymaking in the LDCs (and, indeed, in industrialized countries as well), there are also trends in the direction of responsible social choices that are increasingly evident. There are countervailing forces to redirect discovery and investment toward social goals that transcend the immediate self-interests of the present elites.

Part of the reason for the rise in responsible social choice is that political elites do not always recognize the crucial decisions that determine the future directions of their society; they prefer to leave long-term problems to scientists and technicians. Important social choices are therefore made by international and domestic "experts" concerned with issues that are not perceived by political leaders as central to their interests. Thus, scientists increasingly find themselves able to inject long-term and equity considerations into national policies and economic development plans. The international influences at work on these scientific groups range from statements of human rights and exhortations in the U.N. charter and other documents to specific interventions, by way of research and agriculture, to improve the quality and supply of food. In addition, these scientists have their own links to the international community that draw their attention to the growing concerns over population, nutrition, and ecology expressed by their colleagues in the developed nations.

International fads have not always favored social equity, of course: development planners of the fifties, following contemporary trends, had incorporated into their budgets large-scale, capital-intensive, and socially destructive investments that are still a burden on their nations. Those plans were derived from a now obsolete understanding of modernization derived from the Industrial Revolution and the social theories that emerged thereafter, which were little concerned with equity issues and social values.[13] When the planners express such concern now, they may sound faddish and insincere, but in fact these issues represent permanent and increasingly important problems for scientific research. The influence of social issues on national planning is growing, not waning. More and more national plans are concerned with the quality of life, with the environment, with population, and with basic human needs like nutritional adequacy. The expressions of purpose and the targeted goals of program activity in many less-developed nations are therefore becoming increasingly complex. They call for decisions that go far beyond the "minimalist" choices heretofore considered adequate in science and development policy.

Thus, Praetorian regimes in which scientists are allowed to make such "unimportant" decisions provide some of the best examples of national planning involving major social purposes in the generation and application of knowledge. They may lead nowhere in their immediate political context (who can tell?) but they still provide intellectual stimulation to other countries. Even Nicaragua, torn by civil strife, had formulated a food and nutrition plan in 1977 that linked government programs to scientific research in a comprehensive, integrated program responding to the needs of the poor. The plan identified three elements of the population considered most vulnerable to the effects of malnutrition: (1) children under five who are members of low-income families; (2) pregnant and lactating women; (3) low-income families, especially in rural areas, at risk because of specific nutritional deficiencies like iron, iodine, and Vitamin A. The programs to be undertaken in reaching these target groups were to involve workers in both the public and private sectors of health, agriculture, and education. Resources obtained from international sources, including the United States, and from regional nutrition and agricultural centers, were to be mobilized at the community and national level. The technologies entailed the use of Vitamin A fortification, salt iodization, and iron enrichment, as well as media campaigns and the development and production of appropriate blended food products, the design of integrated rural development programs, and the analysis of the food distribution system. The plan provided for nutrition surveys and recurrent surveillance activities and the evaluation of program impacts on the intended beneficiaries. It was, in short, an integrated assault on the problem, involving the application of all relevant current knowledge, the development of new technologies, the conduct of social and food research, and the coordination of a massive administrative operation.

All of these aspirations were generated at the technical level in Nicaragua, strongly influenced by international planners and the expectation of a substantial development loan. These technical and international actors were crucial: the plan had to be developed and promoted in the face of political indifference to the social values involved. The strategy of the scientists and planners was simple: to demonstrate the coherence and technical and financial feasibility of a balanced approach, and in the process to develop enough bureaucratic and international support to gain the attention and sympathy of political decisionmakers. This political support, of course, did not come easily. Earlier efforts at nutrition programming had been fragmented and unimpressive: a 1969 law requiring the iodization of salt had never been implemented; a 1972 bill to establish a national nutrition council was never passed; a 1973 announcement that food programs subsidized by the United States (PL 480, Title 2) were to be phased out for want of commitment to nutrition improvement was ignored; even the 1975–79 national plan had made only scanty reference to the problems of malnutrition. But mounting interest in the issues of income redistribution and rural development had begun to dominate the discussions in the national ministries, and a sense of jurisdictional competence, together with the possibilities of reinforced, cooperative efforts, encouraged the scientists and technicians to begin the process of constituency-building, to seek international loans, and to develop commitments to a coherent program from the chief lawmakers.[14] Could a dictator use such programs as a claim to legitimacy? That was the gamble the technicians and planners were taking. They lost it in the end, but it is a gamble taken in almost every authoritarian government when men of good will are willing to serve at all.

The process of coalition-building that was followed in Nicaragua was typical of elite-based efforts to build national programs of social choice around new clusters of emergent technologies. It was not a political ploy to preserve a faltering regime. An official paper described the effort as follows: "The Policy is not expected to create support for a National Food and Nutrition Program where none existed before, but rather, to crystallize the diverse public and private support that exists, broaden its base, and facilitate its institutionalization. . . ." Links with private physicians, educators, administrators, local officials, and social leaders would enhance the professional contributions of the scientists and technicians who had developed the plan.[15]

This kind of direct action seems to be an increasingly important preoccupation of scientists and technicians. Even those doing research within narrowly defined fields usually have to work behind the scenes to influence social choices arising from their findings. They have to make choices: whatever the objective of a scientific endeavor, its potential use involves conflicts in human values. Thus the roles of the scientist as discoverer or as technician are seldom isolated: with what other actor or actors should he seek alliance? Security, knowledge, materialism, comfort, self-expression, community, respect, and rectitude—each value arena described earlier in this chapter offers the prospect of coalition to partners in policymaking whose interests may be adverse, or helpful, to those of larger elements in the society:

1. The armorer who serves the military class can threaten the society it is supposed to protect; weapons systems for offense and defense, or for domestic and international uses, elicit different political partnerships.

2. Pure scientists can threaten the establishment with their findings or they can help traditional religious leaders interpret new technologies in accordance with old folkways.

3. Industrial technicians working in a government research institute can concentrate on projects beneficial to large corporate interests or on small-scale projects or on consumer welfare.

4. In the service of medical science, new discoveries may raise the cost of treatment to a point where physicians have to choose which lives to save, or whether or not to impose the burdens of longevity on an unwilling society.

5. By providing or replacing skilled workers with precision tools, technology can help one group of actors in the production of goods and services but only at the cost of injury to others.

6. Which organized group of workers should a technology favor: rural producers (farmers' associations) or the consuming urban proletariat (labor unions)?

7. A technology that displaces the livelihood of a slash-and-burn mountain people also destroys their way of life and self-respect; the technologist has natural allies whatever side he chooses.

8. Even religions contend; a scientific discovery is rarely neutral among them.

Such dilemmas, involving selections among clients and partners, permeate the process of social choice.

Few scientific endeavors escape consequences that some humane scientists would prefer to avoid. These consequences are not the goal of scientific endeavor, but they occur because organized knowledge changes both the ways groups of people do things and the ways they perceive them. The changes are not uniform, and they are not fully predictable. Each sponsor of scientific research—the army, the corporation, the research and development laboratory, the university— seeks to gain from the mastery of new knowledge. They are not by virtue of their sponsorship the only coalition partners scientists seek in issues of social choice. Sponsors may perceive advantages in withholding knowledge as well as using it (even universities may try to keep armies or police investigators or commercial exploiters or competing researchers from gaining access to some knowledge). Thus the scientist cannot escape a role in social choice by efficiently serving a sponsor, even though its role in social change is supposedly benign. By gaining access to knowledge, the scientist assumes responsibilities that go beyond those of the uninformed citizen. Loyal alignment to institutions serving good social ends does not free him from that responsibility.

A convenient definition of the role of scientists in development is the distinction among the functions of generation, diffusion, and application of knowledge.[16] Each of these functions has to be linked to form a triangle of service to social ends, since each is necessary but insufficient in itself as an instrument of man's betterment. But each function can also be diverted away from social purpose; each can be misused if aligned too faithfully with national ends defined by a political leadership; and each can diverge from the common welfare if its orientation to international purposes results in the neglect of domestic priorities. Scientific policy involves the search for an appropriate balance of resources devoted to the generation, application, and diffusion of knowledge in a given setting. Both national and international science communities are aware that effective science policy requires a balancing among the links to foreign and domestic national suppliers and users.[17]

This definition of functions, however, does not afford much guidance to individual scientists confronted with problems of social choice. And scientists need guidance: expertise in atomic physics, ovulation cycles, or weather modification is not often accompanied by social omniscience. The analysis suggested here is a first step in structuring the inquiry into social choices. But even mastering it is not sufficient: a more active role is often necessary. For an appropriately active role, a different set of definitions is necessary. Scientists whose knowledge confers special responsibilities beyond those of the citizen affected by a forthcoming social choice can discharge those responsibilities in one of six ways. They can serve as *advocates* (entering into the policy arena directly), *advisors* (analyzing the consequences of different courses of actions for different groups), *decisionmakers* (in which case they themselves, not their sponsors, accept the responsibility for the uses of their science), *critics* (the advocates of alternate choices); *conspirators* (organizers of opposition forces), or *activists* (leaders of demonstrations and other protest movements against a given use of scientific knowledge).

Dividing one's activity among these roles is the element in social choice that goes beyond minimalist staff work. The calculus of social consequences—whether through technology assessment, utilitarianism, Paretian optimality, or social justice theory—is only the first stage in the exercise of social responsibility. The information derived from this form of staff work is insufficient without the exercise of one of the more active roles. The scientist may not have to play such a role very often, but where his own work is involved, he cannot evade it unless he is willing to allow the product of his hands to serve the immediate self-interest of the strongest and most powerful of his contemporaries.

NOTES TO CHAPTER FOUR

1. Walsh McDermott, "Modern Medicine and the Demographic Disease, Pattern of Overly Traditional Societies: A Technologic Misfit," *Journal of Medical Education*, 41 (supplement) (September 1966): 137—162.

2. John D. Montgomery, "The Challenge of Change," *International Development Review* 9, no. 1 (March 1967).

3. Richard S. Eckaus, "The Factor Proportions Problem in Underdeveloped Areas," *American Economic Review* 14, no. 4 (September 1955); 50, no. 2 (May 1960); A.S. Bhalla, *Technology and Assessment in Industry* (Geneva: ILO, 1975).

4. John D. Montgomery, *Technology and Civic Life: Making and Implementing Development Decisions* (Cambridge, Mass.: MIT Press, 1974), chs. 5 and 6.

5. François Hetman, *Society and the Assessment of Technology* (Paris: 1973).

6. See the paper by Gustav Ranis in this volume and an unpublished paper by A. A. Fouad, "Choice of Technological Models by Developing Countries" (Presented at the Wingspread Symposium on Social Values and Technology Choice in an International Context, June 1978, Racine, Wisc.).

7. Myres S. McDougal, Harold D. Lasswell, and Lung-chu Chen, "Nationality and Human Rights: The Protection of the Individual in External Arenas," *Yale Law Journal* 83 (1974): 900-98; Harold D. Lasswell, *Preview of the Policy Sciences* (New York: Elsevier, 1972).

8. These eight values correspond to a similar list used in the policy sciences: power, enlightenment, wealth, well-being, skill, affection, respect, and rectitude.

9. Alex Inkeles and David H. Smith, *Becoming Modern: Individual Change in Six Developing Countries* (Cambridge, Mass.: Harvard University Press, 1974); Daniel Lerner, *The Passing of a Traditional Society: Modernizing the Middle East* (New York: Free Press of Glencoe, 1964).

10. Montgomery, *Technology and Civic Life*, ch. 3.

11. A few years ago, a group of Harvard scientists identified a geographical site for a nuclear power plant to which no identifiable elements of the community objected. Jacques Gros, *Power Plant Siting: A Paretian Environmental Approach*, Discussion Paper 74-4 (Cambridge, Mass.: Harvard University Environmental Systems Program, August 1974).

12. See my essay "Public Interest in the Ideology of National Development," in *The Public Interest*, ed. Carl J. Friedrich (New York: Atherton, *Nomos*, vol. 5, 1962).

13. "A mutual commitment of both donor nations and developing countries to the Basic Needs model would inevitably have led to a major reorientation of the science and technology production system in the industrial world. So far, [scientists] have largely remained insensitive to the needs and dilemmas of the Third World, be they in the area of appropriate technology, or in the search for development strategies that are respectful of freedom and human rights . . ." Soedjatmoko, "National Policy Implications of the Basic Needs Model" (Jakarta, March 7, 1978). I am indebted to Jonathan Silverstone for this quotation and for other helpful suggestions. He also called my attention to the following sentences in the Franck Report to the Secretary of War (June 1945): "Scientists have often been accused of providing new weapons for the mutual destruction of nations, instead of improving their well-being. It is undoubtedly true that the discovery of flying, for example, has brought more misery than enjoyment and profit to humanity. However, in the past, scientists could disclaim direct responsibility for the use to which mankind had put their disinterested discoveries. We feel compelled to take a more active stand now because the success which we have achieved in the development of nuclear power is fraught with infinitely greater dangers than were all the inventions in the past" (*Bulletin of the Atomic Scientists*, May 1946).

14. The loan was held up in Congress because of Nicaragua's human rights record, then finally approved because of its exceptional merits as an application of technology to a major poverty problem.

15. "Population Policies as Social Experiments" in John D. Montgomery, Joel Migdal, and Harold Lasswell, *Patterns of Policy: Comparative and Longitudinal Studies of Population Events* (New Brunswick, N.J.: Transaction Press, 1979).

16. Jorge A. Sabado, "Science and Technology in the Future Development of Latin America" (Report presented to World Order Models Conference, Bellagio, Italy, 1968).

17. John D. Montgomery, "Science Policy and Development Programs: Organizing Science for Government Action," *World Development* 2, nos. 4 and 5 (April–May 1974).

※ *Part II*

Appropriate Technology in Practice

✳ *Chapter Five*

Appropriate Technology
and the Development Process

Gustav Ranis

The initial task of this chapter must, unhappily, be one of definition. So much has been said and written in recent years about "appropriate," "intermediate," "labor-intensive" or "traditional" technologies—as opposed to "inappropriate," "capital-intensive," and "modern"—that to present one's own interpretation of "appropriateness" before plunging into the subject becomes virtually obligatory. It would be comforting to say that the issue is wholly semantic in nature; unfortunately, it is not. One of the reasons this field is shrouded in such an unusual volume of controversy and confusion is this very lack of substantial agreement on what people are really talking about.

What I shall mean by an appropriate technology choice is the joint selection of processes and products "appropriate" to the maximization of a society's objectives given that society's capabilities. This means that the appropriateness of a technology choice encompasses product quality and technique defined in relation to both the society's endowment and its objectives. In other words, a society's endowment and its preferences as between growth and the satisfaction of (absolute) basic needs and/or of (relative) equity should give us a unique match-up of the "right" basket of techniques and goods known to man—or better, capable of being developed by man.

This seems a tall order but it is really not so complicated, at least not conceptually. An optimal choice can be established if all the information is available and all the known choices are realizable. Our

A paper based on similar materials appears in *Technology and Development: A Realistic Perspective*, ed. S. Rosenblatt (Boulder: Westview Press, 1979).

task is further simplified by the fact, only asserted here, that there need be no conflict between such objectives as growth and equity in developing countries with a labor surplus. In other words, appropriate technology choice can significantly improve performance in several aspects of an economy: employment, income, wage shares, and the family distribution of income.[1]

The real problem is not that our definition of appropriateness is wrong but that it is tautological. The proponents of the concepts "Small is beautiful," "Second-hand is appropriate," and "Labor intensive is good" are able to make much stronger positive statements. Like the "big push" cultists of yesteryear, they know what the proper choices are. *Their* problem is that they may well be wrong. *I* instead have opted for being tautological. What I am saying, to put the case more positively, is that "the" appropriate technology will vary across countries according to differences in endowment and in tastes and within a country over time. Moreover, the appropriate process for a poor labor surplus economy is not always labor intensive, and the appropriate good is not always a basic good. There is even less validity to the presumption that technologies appropriate to developing countries must be somehow "traditional" or, at best, "intermediate" in some sense of progressivity or modernity.

There clearly exists an empirical presumption that the more severe the population pressures on the land, the larger the population, and the greater the shortage of capital, the more likely that appropriate technologies will locate themselves at the labor-intensive and basic needs end of the spectrum. But all the empirical evidence we have accumulated to date leads us to conclude that appropriate technologies are likely to be modern, machine-paced—and drawn from an array of current blueprints—rather than of the traditional, handicraft, or second-hand machinery varieties.

Appropriate technologies can be "advanced," but they need not necessarily be equated with the latest "advanced country" technology. They can be modern *and* labor-intensive, or modern *and* capital-intensive, use imported *or* domestic core technology, make use of extensive local adaptations *or* not. There is no easy, comfortable answer; it "depends"—on the place, the resources, the preferences, and the time. "Appropriateness" is little more than a useful tautology that alerts us to the existence of a wide array of technologies among which the one best suited to the particular circumstances can, at least potentially, be located or devised.

The problem is no longer related, as it was a decade earlier, to the failure to recognize the existence of such a wide diversity of potential choices, particularly with respect to alternative processes. A number of surveys document the existence of a range of alternative factor proportions for all but a small subset of continuous process industries.[2] These surveys agree that there exist substantial substitution possibilities even with respect to the core process. In industries ranging from sugar refining to beer to textile to shoes, there may be as many as five major choices even for the core technology. Moreover, by changing the speed of operations, the number of shifts, the intensity of the maintenance of the machinery, each of these core choices can be made substantially more flexible.

In addition, even greater opportunities for choice exist in the peripheral production activities such as in transporting materials within the factory or storing and packaging the final product. The bulk of the evidence indicates that supervisory labor to manage a large number of the unskilled along a machine-paced production line may be more readily available than the skilled foremen required for highly automated lines. Finally, there is agreement that alternative organizational devices, including taking advantage of economies of scale at some stages of processing while subcontracting other stages, can provide a substantial reduction in the costs of plant, thereby reducing overall capital-intensity.[3]

The new conventional wisdom about the wide range of technology choices is less firm in the realm of the range of alternative goods within a narrow spectrum of Standard International Trade Classifications (SITC). Following the theoretical work by Lancaster[4] and its application to developing countries, especially by Stewart,[5] our increased concern with the appropriateness of goods as a form of technology choice can be analyzed by the decomposition of a commodity into a bundle of quality characteristics, with "planned" as opposed to accidental variations in such characteristics associated with substantial variations in the basic process choices open to the entrepreneur and to society. It is curious that when an economist talks about technology choice he is usually referring to a process choice, a choice of techniques, whereas the businessman is much more frequently concerned with small variations in product choice, a choice of quality attributes. Our basic premise will be that at any level of SITC classification the assumption of homogeneity in the bundle of quality characteristics attaching to a particular good may in fact obscure important residual sources of choice that entrepreneurs can take advantage of, both in utilizing the existing factor endowment and in providing for the so-called basic needs in the internal markets of developing countries.

I do not wish to be misunderstood in one crucial respect. It is not my view that there exists some vast national and international shelf of technology, stretching across all contemporary countries and across recorded time that has on it somewhere just the "right" process and/or product to be plunked down in a particular country and industry context. Although the choice of a technology already in use elsewhere is a crucial step—and by no means easy or costless, as we shall see—modifications will almost always have to be made before it can be installed and become fully "appropriate." Such modifications may constitute major technology changes on "top of" say, imported technology, or minor adaptations or "twists" in what has been or is being used elsewhere. The adaptation of an appropriate process and/or an appropriate good within a developing country thus always requires combining selection from the existing vast array of blueprints with the major or minor modifications required to suit the always slightly different local conditions. Each private or public entrepreneur makes both these choices, often simultaneously, whenever a production decision is contemplated.

However, despite the frequently wide choice of potentially available processes and alternative specifications for commodities, we are still confronted with the empirical fact that the selections actually made in most of the developing countries appear to be substantially "inappropriate" by any quantitative or judgmental standard one can apply. Capital-labor ratios differ between rich and poor countries for given standardly specified goods, but they differ much less than the well known differences in endowments would lead one to expect. Output mixes also differ but not as much as resource gaps and the theory of international trade would predict. Certainly while the per capita consumption of drip-dry shirts and Western shoes in the developing world is lower than that of bush shirts and open-toed sandals, the volume of production and consumption of the latter is less than might be expected.

These are easily substantiated facts that impress the observer. More interesting and more important, of course, is the explanation. A litany of domestic and international causes is usually advanced; they often read like a shopping list of the old fashioned "factors in economic development" or "factors in the selection of appropriate technology." Any serious effort to understand the problem of appropriate technology must order these elements in some meaningful fashion as a preliminary to the development of any internal or external set of policy actions. The obvious and insistent gap between textbook appropriateness and what one actually encounters in the real world, with serious consequences for both employment and income distribution objectives,[6] will be analyzed in this essay as background for the discussion of policy issues. My analysis is divided into two parts: First I shall deal with the effective demand within the developing countries for appropriate technology as I have defined it, and, second, I shall consider the effective supply of choices.

The demand factors center on the conditions at the level of the individual entrepreneur, public or private, which cause him to put a high priority on the search for a more appropriate set of processes and goods. This perspective has usually been the province of economists examining the signals provided by the economic environment for the individual actor, signals that may or may not accurately reflect the basic resource conditions of the society. The supply side focuses on the availability of theoretically obtainable alternatives and the human and institutional search and response capacities that a society needs to discern and develop them. Let us examine obstacles to the effective working of such a technology market, on both the demand and supply side.

THE EFFECTIVE MARKET FOR APPROPRIATE TECHNOLOGY

The effective demand for appropriate technology has at least two dimensions: one is the incentives for individual decisionmakers to seek out the best techniques actually or potentially available; the other is the extent to which their private search for "appropriateness" coincides with or deviates from social "appropriateness." While the first dimension has been relatively neglected, the second has received a lot of attention from economists under the general heading of "the impact of relative price distortions on technology choice." It may nevertheless be true that the first is substantially more important in the real world, certainly in the case of nonagricultural activities.

The typical developing country, as most readers will appreciate, emerged out of the colonial production and resource flow structure in the years after World War II (somewhat earlier in the case of Latin America). The postindependence policies were characterized by an increase in protection and intervention by national governments. Specifically, instead of allowing receipts from primary exports to be used for reinvestment in traditional sectors and in service activities facilitating these sectors, a well-known "bundle" of primary import substitution policies was instituted. The purpose was to shift both foreign and domestic resources into the rapidly growing consumer goods industries supplying the domestic market. When this particular subphase of development ran out of steam, most of the countries turned, under a continued and possibly even stronger version of the same policy, to what is usually called secondary import substitution: the replacement of previously imported capital goods and the processing of raw materials. Only a minority of countries, such as Taiwan, took a different path at the end of primary import substitution—one of continuing to produce but not to export the same relatively more labor-intensive consumer goods previously supplied only to the domestic markets.

The basic point here is that the policies of import substitution, primary and even more so secondary, are by their very nature intended to provide protection to the fledgling and, later, not so fledgling, domestic industrial producers. This translates into a dispensation of all kinds of noncompetitive favors, not just with respect to protection from foreign competition but also in terms of providing an oligopolistic or monopolistic situation in the domestic industrial sector. Windfall profits are thus guaranteed through the import licensing system, with overvalued exchange rates, artificially low interest rates to favorite borrowers, the direct allocation of strategic materials such as steel, cement, and so forth, and the entire panoply of by now well-known favors, all aimed at ensuring safety and profits to the new class of entrepreneurs.

There is no need for one more critique of the import substitution policy syndrome and its effects. The point to be made here is that it leads to a lack of competitive pressure on entrepreneurs, especially in the nonagricultural sector. When profits at the level of 25 percent or 30 percent are guaranteed individuals simply by virtue of their being in front of the queue for whatever "goodies" are being allocated, the desire or felt need to seek out the most appropriate technology is severely blunted. Although it is not part of the profit-maximizing religion of economists, it is nevertheless observable that entrepreneurs are much less anxious to increase their profits from 20 percent to 30 percent than from 10 percent to 15 percent, assuming that they have been guaranteed their basic cushion by virtue of the monopoly windfalls granted by government action. Such satisficing behavior or preference for the "quiet life," as it is sometimes called, is a phenomenon observed in advanced countries as well.

The prevalence of satisficing behavior of this kind—when individuals' energies are much more profitably spent seeking government favors and ensuring the maintenance of their monopolistic positions than searching for more appropriate goods and processes—cannot be underestimated. The reason that it has taken economists so long to give this dimension its proper weight is, I believe, the fact that it is relatively less important in agriculture, which remains the dominant activity in most developing economies. The usually small-scale, atomistic nature of the agricultural sector in developing societies makes this particular route to inducing an inappropriate technology choice less important. The typical farmer is normally "up against it." He is highly averse to risks and depends on the government to convince him that new agricultural technology—more often produced in the public sector—is something he should be interested in. It is not the exercise of his market power that leads him to reduce his effective demand for appropriate technology.

Let us turn now to the second dimension of the impact of the import substitution syndrome on the effective demand for appropriate technology. The basic proposition here is that the thicker the veil between an economy's equilibrium or shadow prices for factors of production and the processes that appear in the market as a consequence of various kinds of government interventions, the more distorted the technologies that actually emerge from the selection process. The frequent overvaluation of the exchange rate cheapens capital goods and imports. Minimum wage legislation, protection of workers against dismissal, the encouragement of unionization, and a long list of social legislation artificially raise the price of unskilled labor. Low-interest policies and accelerated depreciation allowances bias entrepreneurs in favor of the use of capital. Price controls on basic consumer goods as well as such important producer goods as cement and steel lead to distortions in terms of quality as well as technology choice. The list is considerably longer. Moreover, the distortions seem to be pronounced for larger firms, for example in terms of the relative impact of credit rationing and of the reach of minimum wage legislation. The degree to which such distortions are more severe in some countries than others depends on the state of inflationary pressures, the relative flexibility of the protective measures (for example, between the use of tariffs and quantitative controls) and the differing impacts of protective regimes on the technologies involved.

One important and little-studied impact of the import substitution policies on effective demand for appropriateness in technology is the discouragement of the production of appropriate goods as a consequence of the relative discrimination against agricultural output. I have already noted the usual effect on agriculture's terms of trade but have not commented on the fact that the resulting lowering of agricultural incomes, coupled with concentration on the substitution of domestic imitations for previously imported consumer goods, has the serious consequence of stifling the demand for more appropriate, basic goods.

As an example, in the Philippines during the colonial period, imported consumer goods managed to kill off much of the efficient handicraft industry in the outlying islands. When import substitution took over in the post-war era, textile manufacturers in Luzon were in a position to "kill off" what little demand for handicrafts there might have been in the outlying areas of Mindanao. The big problem, of course, was the lack of market demand that occurs when rural growth does not keep pace with urban growth. This is in addition to the fact that the existing demand was being absorbed by inappropriate or overspecified urban goods.

Since there exists a strong underlying demand for more consumer goods in developing countries, there is no lack of effective demand, in the advanced economy sense, for final goods. What is needed is a growth strategy that is sufficiently participatory in character to satisfy basic demands across the vast landscape of the typical developing country. The availability of appropriate goods as well as appropriate processes that together provide the "utils" consumers are seeking and also permit a greater utilization of the typical developing economy's abundant resource, that is, unskilled labor, is essential. The narrowly focused, urban-oriented import substitution path to growth has as one of its consequences the suppression of a potential domestic market for appropriate goods. Thus, still on the demand side, entrepreneurs are likely to be fixated on the demands of the industrial urban elite as well as, in some cases, on the demands of export markets. Once again, the effective demand for appropriate technology, as I have defined it, is weak as a consequence of the growth path chosen. Given a relatively stagnant agricultural sector, we cannot expect rural industries and services to grow, whether they constitute inputs into the agricultural sector or outputs demanded by rural populations over time. Bamboo tube wells are an example of the first category, and locally manufactured textiles fall into the second. The spectacular growth of rural industries around Daska in the wake of the agricultural spurt generated by Pakistan's Green Revolution of the sixties provides a historical case in point.

Working in the context of an advanced economy, Schmookler has presented the thesis that demand is the main determinant of invention.[7] He found that the number of inventions in a given field of capital goods tended to vary over time with the sale of capital goods and, further, that the number of inventions, as measured by patents, tended to vary among industries in proportion to their share of the sales of capital goods. If supply factors can be neglected, it followed for him that inventive activity would be greatest where prospective sales and profits were greatest. Schmookler was talking about final consumer demand rather than the demand for technology, but they are related, especially in the area of the appropriateness of different kinds of goods. The stagnation of rural incomes, for example, serves to stifle market demand and thereby the demand for appropriate kinds of technologies.

In summary, we may say that both the level of overall competitiveness in the developing society and the specific impact of policies that distort signals between factors and commodities, as well as for whole sectors, have the effect of inhibiting the demand for appropriate technology and may be viewed as a serious obstacle to making appropriate choices. Removing such obstacles, by the same token, is a necessary but not sufficient condition for achieving appropriate choices. As Rosenberg points out, "economic forces and motives do not act within a vacuum but within changing limits and constraints of scientific and technical knowledge."[8] If the required knowledge with respect to alternative processes or quality bundles is not available and/or the capacity to modify what is on the shelf does not exist, appropriate technology cannot be selected nor can innovations of greater appropriateness be made.

To complete our story of the effectiveness of the market for appropriate technology in the developing world, we must now turn to the supply side. If we picture the full storehouse of human knowledge potentially available to a developing society today, we would, of course, include every known combination of factors needed to produce a given good, plus every known alteration of a quality bundle. This gives us a potentially huge "shelf" of production points. An immediate problem is that such a shelf really does not exist. First, there is frequently a lack of knowledge with respect to a technology that may have been in use, say, twenty years ago or even in use today in another country or in another part of the same country. This means that there is a remarkable lack of information in the hands of the individual decision-maker. Put differently, there are high search costs involved in illuminating a wide range of points along the contours of production possibilities. Second, in practical terms, some of the needed information is unknown because of imperfect competition on the side of the technology-seller. (This applies to both processes and goods and more strongly to product or quality differentiation.) Given the constraints of patents, licensing, trademarks and so on, a good deal of the information with respect to potential technology choices is not readily available to all seekers. This situation is independent of the additional problem that technology itself may not be institutionally separable from other parts of the capital transfer bundle, as, for example, in the case of a multinational investment that comprises capital, technology, management, and technical assistance.

Within the recipient developing country, there must exist, moreover, both the human and the institutional capacity needed to make appropriate choices. That is to say, even if the demand conditions are approximately "right," there is a question of being able to discern, from the full storehouse of technological information, the proper combination to fit the particular endowment. We hear all too often of the neglect of "third factors" and overheads that make the transplantation of particular pieces of foreign technology highly inappropriate. At best the attempt to recreate a foreign production function within a developing economy may lead to a substantial period of so-called x-inefficiency.[9] At worst, cases of "turnkey projectitis" lead to permanent inefficiency and inappropriateness of the process or product. It should be emphasized again that this represents both a human and an institutional failing on the supply side as opposed to the imperfect competitive and market-distorting features on the demand side—although admittedly these conditions may occur simultaneously and be difficult to disentangle in any real world situation.

The institutional capacity involved here entails not only the ability to make the right choice, but also the ability to diffuse a technology once it has been selected. Diffusion channels through institutes, salesmen, the private banking system, government extension services, catalogs, fairs, and other channels is not a trivial matter. These networks do not always exist in many developing countries or they may be strongly biased or inefficient. I have observed, for example, the use of a highly efficient bamboo tube well in one part of Bangladesh not known to similar regions in other parts of the same country.

This problem represents, in part, an extension of the aforementioned information problem to particular kinds of identifiable institutional handicaps. The ability to ask the right questions, not in terms of international reference services or lists of technology alternatives, but in knowing one's own situation specifically enough to request what is really needed is not common. It requires, for example, a minimum level of scientific and technological capacity, not necessarily based on high levels of educational attainment but on the kind of technical literacy that many societies do not yet have. And, as I shall note below, many of the science and technology institutes one finds in the developing countries are too often concerned with other matters.

There are further supply constraints that should be mentioned. We have already, in passing, referred to the monopoly power frequently exercised by the multinational corporation. This is especially visible in the case of product differentiation in consumer goods where technology is substantially appropriable and where the imported technology may be subject to market sharing arrangements and to the response of the multinational corporation to its own internal competitive pressures. Furthermore, all the supply bottlenecks are by no means indigenous to the developing country; many have their origin in the rest of the world with which the typical developing country has to deal, whether or not they are part of the bundle involving the capital good import proper. This includes both aid tying and inflexible rules of multinational behavior.

Finally, in this observer's view at least, the most important obstacle to the availability of an effective supply of knowledge for the developing society is the recognition that no piece of technology, wherever found, is ever quite "right" and ready to be put into place in some given local context. While differences in the resource and other environmental conditions may be minor, especially as we go from one region of the same country to another, I am very much impressed with the importance of human and natural "third factors," which can never be fully taken into account in any engineering production function. As a consequence, the ability to make minor adaptations or "twists" on what already exists somewhere in the storehouse of human knowledge is of extreme importance. Such capacity may be found in the repair shops of the large company, in research and development institutes, or in the much less romantic confines in which the village blacksmith or the bicycle repairman work. Without a technical problem-solving capacity, often derived from machinery maintenance and repair functions, modernized local versions of secondhand machinery or appropriately modified foreign imports cannot be expected to appear. With it, the range of potential usable points in the production possibility set is substantially expanded.

We are aware, of course, that selection and adaptation often go hand in hand, but it is important to differentiate between them conceptually. It is also necessary to recognize that the human and institutional capacities to make choices from a static array of possible technologies and the capacity to carry out these sometimes minor twists and adaptations are not necessarily identical.

The capacity to obtain appropriate technology by selection and adaptation is related in very complicated ways to a country's scientific literacy. As has been pointed out by Kuznets and others, the causal chain does not always run from science, which is more of a universal good, to technology, which is more of a national or even subnational good, but may run from a problem that needs to be solved at the technological level back to science. The experience of international agricultural research in recent years via the so-called Green Revolution is very instructive. We all know about these substantial agricultural advances based on genetic research, for example, the so-called dwarf varieties of rice developed at the International Rice Research Institute at Los Baños. Yet it has become increasingly clear in recent years[10] that without a national capacity for technological adaptation, the Green Revolution would have run into serious difficulties by now—the moisture, soil, pest, and temperature conditions never being quite the same in two given situations. Similar problems apply to industry, for example, the humidity conditions affecting leatherware and the quality of the water supply affecting plastics.

STRENGTHENING THE MARKET FOR APPROPRIATE TECHNOLOGY

If the preceding diagnosis of the problem is, in the main, correct, certain policy conclusions follow. Perhaps the most telling, certainly for someone viewing the problem from abroad, is the extent to which the strengthening of both the demand and the supply sides of the market for appropriate technology depends on decisions within the developing world. The venerable notion that technology transfer constitutes the main instrument available to the technologically advanced countries of the world to help the poor quickly, effectively, and (even better) cheaply, is quite off the mark, at least in its simple form. Whether we are talking about public transfers of technology of the Point Four variety, or private purchases in the open market, selection of appropriate technologies unfortunately does not come easily. Leaders in technology can be helpful but not without a prior understanding of the basic obstacles to a better functioning of technology markets. In the absence of such understanding, there is little doubt that even well-intended outside help is likely to contribute more to the problem than to its solution.

What then is the proper view from abroad? Not surprisingly, it is one that seeks to identify those policies and actions less-developed countries require to remove, or at least reduce, the obstacles in the way of, first, an effective demand for and, second, an effective supply of appropriate technologies. Only in that context is it sensible to ask how we can really be helpful, if we are so asked.

Turning first to the demand side, this is most clearly a function of the LDC government's macroeconomic policies. The greater the flexibility of the virtually inevitable import substitution regime and the shorter its life, the less severe the costs of misplaced entrepreneurial attention and initiative. The greater the domestic freedom of entry into industry, the less serious the impact of the "quiet life" financed by windfall profits. And, of course, that holds as well for exclusive market arrangements and no-export clauses asked for, and bestowed upon, transnational corporations. In brief, the more "workable" the competition in this phase of development, the lower the costs in terms of the foregone active search for appropriate technologies.

The same arguments hold for the extent of relative price distortions in factor and product markets. Policy regimes that gradually liberalize and move toward more market-oriented export substitution solutions have a much better record in choosing appropriate technology than those that retain an increasingly heated import substitution hot-house.[11] The existence of a heavy veil between relative endowments and factor prices, and between the availability of various types of intermediate and final goods and the administered prices established for them cannot help but have the expected negative effect on the selection of appropriate technologies. Regardless of whether appropriate technologies are selected, we find that the more severe the distortions of the terms of trade against agriculture, the more serious the lack of purchasing power in the rural areas for more appropriate industrial goods. Reducing the magnitude of that veil, via exchange rate and interest rate reforms as well as the liberalization of other markets, is bound to be helpful. The same types of policy reform that make for greater competitiveness and smaller windfall profits can generally be counted on for beneficial impacts here.

The fact that relative prices do matter for the demand side has by now become an almost embarrassing part of the conventional wisdom, but that does not make it any less true. While the precise package of macropolicy changes will of course differ, depending on the nature, scope, and severity of the existing controls system, there is little hope that, in the absence of a vigorous and properly directed demand for appropriate technology, any force-fed attack via increased supply can work. Nor can policy changes be forced by pressure, exhortation, or laboratory demonstrations on the supply side.[12] In the typical developing country, if the millions of individual decisionmakers do not "have the bit in their teeth," little else can follow. This necessary but not sufficient quality of the demand side of appropriate technology markets is well captured by Timmer's *bon mot* that "getting relative prices right is not the end of development, but getting them wrong usually is."[13]

Strengthening the demand for appropriate technology is thus largely a matter for domestic decisionmaking and domestic action.

Foreigners can help by providing advice and technical assistance as well as by providing capital resources to ease the country through a policy reform period, either in real or in psychological terms. By the same token any substantial foreign aid involvement is unlikely ever to be neutral with respect to the recipient's set of policies, since it helps either to maintain those policies or to change them. It is more likely to be a question of the way in which foreign influence is brought to bear, that is, the style of "consultation" or "intervention," rather than its absence or presence.

In a somewhat lower key, foreign private investment can have a similar impact on the macrodecisions LDC governments make or fail to make. Inserting export prohibitions or exclusive rights into contracts does not help; speaking out for the liberal export-oriented policies often preached in the industrialized nations might. But we cannot, of course, expect a transnational corporation to look far beyond its own profit maximizing calculus, which will necessarily adjust itself to the conditions encountered—and projected—in the particular country's situation. In this sense at least, the large foreign subsidiary is not much different from the large domestic firm, although global maximization may call for a shorter exposure in the LDC, given the higher real or imagined risks the transnational corporation thinks it is confronting. Further, a relatively greater preference for governmental policy stability indicates that developed country governments can help alleviate these fears by moving beyond the application of narrow financial criteria when public funds are used to provide subsidized investment guarantees and other favors, via the Overseas Private Investment Corporation (OPIC) for instance.

Most of the studies of this question undertaken at Yale[14] and elsewhere[15] confirm the fact that entrepreneurs are indeed sensitive to price signals and do respond to the relative pressures of a more or less competitive environment with a more or less assiduous search among alternative processes and attribute bundles. But much less attention has thus far been given to the supply side of the equation, certainly by economists who have traditionally tended to leave this dimension to technologists, scientists and transnational corporation executives. Unfortunately, in this LDC market, supply does not create its own demand, nor does demand automatically elicit the required increase in supply.

Turning, finally, to the implications for policy of this assessment of the obstacles inhibiting the effective supply of appropriate technologies, we again find it important to differentiate between the domestic and international actions essential to permit an effective supply response to any reasonably "unleashed" forces of demand.

Thus, I will, where relevant, distinguish between domestic and international policy options.

Let us begin with the possibility of extending the range of information available for existing technology choices. At present the information actually at the disposal of individual economic agents is often severely limited. This is due not just to the inadequacy of the book of international blueprints available to the typical LDC industrialist; more telling is the failure of similarly situated LDCs to share information. This deficiency also occurs among regions and even villages within the same country. The example of the bamboo tube well example illustrates how often it has become necessary to reinvent the wheel, or do without. Such problems obtain in both agriculture and industry, but their solution via government research institutes and extension services is further along on the agricultural side. No equivalent industrial research and development and extension service exists within most developing countries. The problem is not, we believe, tackled effectively by concentrating unduly on international "question and answer" services. The basic information networks within the developing countries themselves must be strengthened; and the same networks that carry information about alternative processes and goods must also have the research and development capacity to help individuals effect the needed twists and adaptations.

Before leaving the information dimension of the question, an additional point needs to be emphasized, namely the impact on the supply side of imperfect competition in international markets, both with respect to processes and (especially) with respect to products—as exemplified by patents, trademarks, licenses, and the like. Although some protection of an advanced nation's innovative developments is surely valid, the present extent of that protection in LDC markets can only be termed unwarranted. In this sense, international patent legislation could be usefully altered to ensure that patents are actually used (and technology transferred); too often patents merely serve as a way of controlling markets and inhibiting the flow of information. Current suggestions for "unbundling" various transnational corporation packages to permit a more realistic and arm's length type of transaction with respect to technology transfer could also be very helpful.[16] It is also possible that transnational corporations could be induced to exchange information on their experience with adapting appropriate technology in their subsidiaries. This could be done on a basis that would not violate sensitive proprietary information but would instead emphasize principles, thus giving the transnational firms a chance to act as catalysts and to make a social contribution in an area where they are often criticized.

LDC governments, through a revision of often lax procedures for monitoring and disseminating information on technology transfer, licensing, and patent agreements, can substantially speed up the diffusion of new technology across firms. For example, the shift in nineteenth-century Japan from mule to ring technology for spinning was almost instantaneous. When economies of scale and information channels are important, as they often are, the encouragement of trade associations on the historical model of the Japan Cotton Spinners' Association, or even the encouragement of trading companies on the more recent Japanese model, should be considered.

By limiting the role of private capital, public aid programs may have a restrictive influence on the range of information actually available to an LDC industrial entrepreneur. This should not be interpreted as a plea for the establishment of a perfectly competitive international trading community with perfect information—a condition that would be highly unrealistic. It is, rather, an affirmation of the fact that some of the mystery and secretiveness surrounding the transfer of technology is often not warranted by the basic economics of the situation but is, rather, the consequence of artificial and often misguided interventions by government and of the unduly proprietary behavior of private interests. Aid, for example, is often doubly tied, once because of the decision by the country of origin to restrict the source of imported technology choice, and again because of its unwillingness to incur local costs, resulting in eliminating the recipient country's own capital goods industry from supply consideration.

Aid agencies should be forced to consider real world alternatives in their project lending rather than be permitted to follow "best" engineering practice after paying lip service to "appropriate" technology choices. *Ceteris paribus*, program lending is to be preferred to project lending, while the willingness to permit local financing on projects will help reduce built-in preferences for capital imports. On the private capital side, a similar willingness to sensitize investors' technology behavior, in return for granting subsidized OPIC insurance and/or contributing public funds to the investment itself, would be among the actions that could be taken from the outside to assist in broadening the effective range of information and choice.

Closely allied to all this but more fundamentally important in my opinion are actions to enhance the developing countries' own capacities for technology adaptation. As pointed out earlier, it is very unlikely that, even given perfect information, a particular process or quality bundle attached to a commodity will ever be precisely "right" for a particular rural industry or a particular local market. The network that will effectively carry information must, at the same time, have the capacity to adapt shelf technology, whether foreign or domestic, to particular local specifications. A combined information, adaptation, and dissemination network is essential because information about, and adaptation of, existing technology cannot and should not be separated. Centers of excellence in particular major industries, in conjunction with multipurpose centers of local adaptation and dissemination, are likely to be needed to ensure a national capacity for "tinkering" in the direction of more appropriate technology. The whole system should be as closely related to the private market as possible. In most developing countries, this might mean a greater role for rural commercial branch banks, or development banks with a technical assistance capacity built in, rather than the independent creation of the industrial equivalent of an agricultural extension service.

The likely costs of such a network in a developing country are not all that large. What is even more encouraging is that few additional resources are probably necessary. Most developing societies already have a substantial set of institutions for the advancement of science and technology. Typically, however, such institutions have shown only a marginal interest in the area of appropriate goods or appropriate processes. Though the total research and development effort is pitifully small, it is largely spent on breakthrough technologies and "frontiers of science" activities patterned on advanced country efforts. The redeployment of these resources in the direction of diffusing information on the feasibility of alternative combinations of attributes as well as on experiments involving more appropriate technologies would be of the greatest value. One way to guarantee that reconstituted institutes of this kind become part of the network I have described is first, to ensure that while they may well require government subsidies at the outset, over time they increasingly pay their own way by means of private sector contracts; and, second, to reward the staff of such institutes on the basis of success in these objectives rather than exclusively by applause from the "invisible college" of fellow engineers and scientists. Assuming that the proper demand conditions exist, it would take very little to shift some of these institutions into the kind of networks I have described.

The appropriate research and development would include both hardware prototypes, such as those focused on the scaling down of advanced technologies, and such software activities as are required for in-plant floor-level technology adaptation. A particular target of attention would be the creation of a capital goods capacity, especially at the village level. Such capacity already exists in most developing countries in the form of repair shops, foundries, and so forth.

It is in such individually non-spectacular, but in toto quantitatively impressive places that indigenous innovative capacity can be found and could be substantially enhanced with a minimum of additional outside inputs. For example, in Sri Lanka industrial cooperatives aided village blacksmiths.[17] Elsewhere, the appropriate linkages can be forged with the commercial banking system or preferably with rural development banks, in contrast to separate rural agricultural and industrial bank structures. Encouragement of a simple capital goods industry within the developing countries is likely to be helpful not only because the industry itself is normally labor-intensive but also because it provides additional technological choices within each of the customer industries.[18]

The emphasis, I believe, should be on the establishment of a functioning network and a decentralized capability to sustain it rather than on the transfer of hardware. This requires the evolution of a strategy of spatial expansion of industries, especially in activities not subject to pronounced economies of scale, keeping in mind the extent and nature of the transportation systems. Such an approach makes sense for both market and labor supply reasons. The reallocation of governmental overhead investments toward the rural areas, thus assisting rather than inhibiting the spread of internal markets, can provide a major stimulus to the development and diffusion of appropriate technologies and appropriate goods.

Building on traditions of local appropriate goods and technologies may seem rather primitive but in many instances it really means revitalizing village industries destroyed by government policies that have favored large-scale urban industry during import substitution regimes. I am not arguing for the preservation of outdated handicraft industries but for the recognition that such industries may well provide in many, though not all, cases the basis for modernized, efficient production activities.

One of the real, often unreckoned, casualties of a society that depends heavily on the importation of hardware and its emplacement without much modification is the consequent lack of participation and learning by experimentation. A decentralized indigenous innovation and adaptation network has the salutory effect of building confidence in a society's own capacity to make appropriate technology choices.

With respect to the more urban-oriented medium and large-scale industries, the same plea for decentralization carries. In this instance, however, empirical evidence indicates that many of the adaptive technology changes that have occurred in the more successful developing countries (Japan, Korea, and Taiwan) have emanated from the repair shops or the factory floor rather than from the research and development establishment as such. Here again it is more a question of systems and software, including the sociology of the firm and its ability to reward suggestions coming up the line, than it is of specific kinds of imported hardware decisions. For a multinational subsidiary in a less-developed country, adaptive change often means overcoming the rules of thumb set by headquarters. The obstacle appears to be more easily overcome in the case of a company that has a number of foreign subsidiaries and thus more experience across differentiated resource environments with what constitutes appropriate technology. The sharing of management organizational devices and experiences in this area among firms and countries without divulging proprietary secrets could certainly be helpful, and incidentally, could serve to improve the sometimes rather tarnished images of the multinationals.

Developed countries can help to strengthen the capacity to make technological choices within developing countries by providing catalytic inputs of the financial and human resource variety for the international network that has been outlined. Concentration on helping to develop an enhanced capacity to ask the right questions and to choose more appropriately would be a wholesome change in emphasis in foreign assistance efforts.

A closely related area for international action begins with the incontrovertible fact that less-developed countries may have more to learn from each other in the area of appropriate technology than from the more advanced countries. The extension of the argument is that some advanced countries, such as Japan, have more to offer than others, such as the Soviet Union and the United States. One policy implication of this is the encouragement of national centers of excellence for the LDCs with a strong regional dimension built in and possibly subsidized initially from the outside. This is perhaps a better way to proceed than to try to imitate on the industrial front what has been successfully accomplished in the agricultural field through the various international crop research institutes. Advice, encouragement, and seed money can often be very helpful in building a regional dimension into national centers of excellence that have been developed as elements of the local networks previously described.

In sum, this analysis gives us grounds for considerable optimism as to what can reasonably be accomplished by policy actions, both internal and external, to render technology choices more appropriate. The concept is clearly multidimensional and the effort to search for simplistic or emotional solutions is likely to mislead us. Neither the return to Ghandian handicrafts nor the search for the "big" technological breakthrough is a particularly useful image to capture the essence of the problem. The center of gravity of the appropriate technology concept rests, rather, with thousands of nonspectacular, adaptive responses and modifications of modern processes and goods across a vast range of applications and landscapes. The latest technology is not invariably inappropriate nor the most basic good invariably appropriate. A capital-intensive technology may be the most efficient, and the poor people do not necessarily wish to buy "poor people's" goods. Cultural imperialism can alter taste preferences in techniques and commodities just as much as a policy of national self-reliance. And governments can always be expected to affect choices—for example, against luxury goods—via tax policy. I certainly am not wise enough to preach on the exact nature of the socially optimal choices. It is the citizens of developing nations who should be given the opportunity of choosing among alternatives, with the fullest possible information and with relative prices more adequately reflecting variations in the quality bundle and with pressures of workable competition on producers. It is only in this way that we can expect to move toward more appropriate technology choices, given a society's resource endowment and development objectives.

NOTES TO CHAPTER FIVE

1. The interested reader is referred to G. Ranis, "Development and The Distribution of Income: Some Counterevidence," *Challenge* (September–October 1977); and J. Fei, G. Ranis, and S. Kuo, "Growth and the Family Distribution of Income by Factor Components," *Quarterly Journal of Economics* (February 1978).

2. D. Morawetz, "Employment Implications of Industrialization in Developing Countries," *Economic Journal* (September 1974); S. Acharya, "Fiscal/ Financial Intervention, Factor Prices and Factor Proportions: A Review of Issues," IBRD, Staff Working Paper 183, 1974; A. Bhalla, *Technology and Employment in Industry*, ILO, 1975; G. Ranis, "Industrial Technology Choice and Employment: A Review of Developing Country Evidence," *Interciencia* 2, no. 1 (1977); Frances Stewart, "Technology and Employment in LDC's" in *Employment in Developing Nations*, ed. Edgar D. Edwards (New York: Columbia University Press, 1974).

3. The empirical reality behind these three loci of technology is discussed extensively in the author's "Industrial Sector Labor Absorption," *Economic Development and Cultural Change* (April 1973).

4. K. Lancaster, "A New Approach to Consumer Theory," *Journal of Political Economy* (April 1966).

5. F. Stewart, *Technology and Underdevelopment* (New York: Macmillan, 1977).

6. These issues cannot be discussed here. See, however, D. Morawetz, "Employment Implications of Industrialization in Developing Countries: A Survey,"

Economic Journal 84 (1974); H. Chenery et al., *Redistribution with Growth* (Oxford: Oxford University Press, 1974).

7. Jacob Schmookler, *Invention and Economic Growth* (Cambridge, Mass.: Harvard University Press, 1966).

8. Nathan Rosenberg, "Science, Invention and Economic Growth," *Economic Journal* (March 1974).

9. A concept based on the notion that firm efficiency cannot be determined by conventional production function analysis only, but that such "third factors" as employee experience, which evade quantification, may be relevant. Such factors, relevant to the transfer of turnkey projects, have been labeled x-efficiency by some economists, following Leibenstein.

10. Robert Evenson and Yoav Kislev, *Agricultural Research and Productivity* (New Haven, Conn.: Yale University Press, 1975).

11. For the contrasting experience of two developing countries, see the author's "Appropriate Technology in the Dual Economy: Reflections on Philippine and Taiwan Experience," in *Appropriate Technology*, comp. International Economic Association (New York: Macmillan, 1979).

12. Any more than (as we now understand) farmers will seek out new seeds if food prices are out of kilter, or families will demand contraceptives if their condition motivates them to have large families.

13. C. Peter Timmer, "Estimating Rice Consumption," *Bulletin of Indonesian Economic Studies* 8, no. 2 (1971).

14. M.A. Bailey, "Technology Choice in the Brick and Men's Leather Shoe Industries in Columbia"; Lucy A. Cardwell, "Technology Choice in the Men's Leather Shoe and Cotton Spinning Industries in Brazil: The Relation between Size, Efficiency and Profitability"; John Fei, "Technology in a Developing Country: The Case of Taiwan"; Gustav Ranis and Gary Saxonhouse, "Technology Choice, Adaptation and the Quality Dimension in the Japanese Cotton Textile Industry."

15. For example, Howard Pack, "Policies to Encourage the Use of Intermediate Technology" (mimeographed report for AID, April 1976).

16. For more on the potentially helpful role of the transnational corporation, see the author's "The Multinational Corporation as an Instrument of Development," in *The Multinational Corporation and Social Change*, eds. Louis Goodman and David Apter (New York: Praeger, 1976).

17. Nicolas Jéquier, ed., *Appropriate Technology—Problems and Promises* (Paris: Organization for Economic Cooperation and Development, 1976).

18. For more discussion on the importance of the capital goods industries, see Howard Pack and Michael Todaro, "Technical Transfer, Labour Absorption and Economic Development," *Oxford Economic Papers* 21 (November 1969).

The Role of High Technology
in the Industrialization of Korea

Hyung-Sup Choi

Technology fulfills many roles in today's societies. In developing nations, technology has come to be viewed as one of the most important means of achieving the aim of national development. This chapter considers the role that science and technology have played in the development of Korea. Particular attention will be directed toward the use of high technology, for this was the path chosen by Korea in its press to industrialize and to evolve an outward-oriented economy. Although less-sophisticated technologies can surely serve the needs of some aspects of national development, Korea determined that the high-technology path could afford her the most options in reaching development goals.

In presenting the Korean experience, I shall be giving more attention to the role of technology in national development programs, not because I believe in a highly centralized system but because of the impact technology can have at the national level. The national government has a crucial role to play during the various stages of development with respect to the goals of development and the choice of technology selected for achieving the ends. Korea's decision to choose the high-technology path resulted from the realization that the problems that burden a developing country need a bold and innovative strategy for their elimination.

In the three decades since World War II, many countries have made determined efforts to overcome the vicious cycle of underdevelopment. Most have sought to industrialize their economies, thereby reflecting the aspirations of their people. Such decisions, for obvious reasons, need to be based on a clear understanding of the potential the country possesses and the constraints to which it is subjected. If it is richly endowed with natural resources for industrialization, the approach must be different from the case in which there are rich human resources, as in Korea, but few natural ones.

Let us examine what prompts many developing countries to opt for industrialization. First, industrialization is most usually a high-priority concern for economic planners in developing countries because of the need to:

1. Create employment
2. Meet domestic demand for products that otherwise would be imported
3. Earn foreign exchange by producing goods that can be sold on world markets

The industrialization strategy of most less-developed countries may be characterized as import-substitution—domestic production, under tariff protection, of goods that were formerly imported. A protected seller's market provides little incentive to apply technological innovation for either market penetration or cost-saving production processes, and, as a result, the industrialization strategy adopted by many less-developed countries is partly responsible for their failure to develop their own research and development systems. A noncompetitive market generates little incentive for the economies and advances that science and technology can contribute.

Some countries have, however, coupled export-promotion with import-substitution, and in these countries the need to compete in world markets generates interest in strengthening the technological capacities of local enterprises. When the less-developed country chooses the industrialization pattern, it must aspire to more than the employment of low-wage, low-skilled labor for simple assembly and shop operations. It must expand its industrial base, upgrade labor skills, and modernize the manufacturing system.

Most developing countries suffer from a vicious cycle of "underdevelopment" of many kinds, but particularly in the economic and institutional sphere. Thus, in our experience it is imperative to have some lead sectors developed with a daring mix of advanced technologies relevant to the individual country's absorptive capacity and entrepreneurship, which can be nurtured in the environment of a free economic system. Here lies an area for the interplay of politics and economics, which, if properly arranged, can result in a synergistic effect. Almost total reliance on imported technology, without a corresponding industrial research capacity, has often meant that most of the technology developing countries receive has been ill-suited to their needs and conditions. Either product designs or production methods or both have been inappropriate.

An adequate understanding of the process of modernization and economic development presupposes an analysis of its goal structure at the national level. The term "goal" refers to aims that are consciously being pursued by those in the polity who make major decisions. The Korean government has repeatedly declared that it has three basic goals: first, national security, followed by economic growth and national solidarity. These three goals are listed clearly in order of priority: national security is the ultimate value and economic growth and national solidarity are the instrumental values needed to achieve the primary goal. Once we understand the overall goal then we can see where and what type of technology should be used in the development effort.

Science and technology when applied with care and thought can help to eliminate the obstacles to national development, for, in some sense, industrialization is the offspring of modern technology. There is a tendency, however, to advocate optimization of the factor endowments of a given country; this results in both the deployment of less-sophisticated technology and an emphasis on importing from abroad whatever advanced technology is needed. The mere assessment of the appropriateness of technology uses conventional yardsticks to take various other important factors into consideration, for instance, absorptive capacity, which itself depends on a myriad of factors such as the institutional, legal, cultural, and environmental system of the country. Thus, labor-intensive, capital-saving technologies—collectively known as appropriate, alternative, or intermediate technologies—may be suitable for a developing nation that strives to build up industries linked with both domestic and outside markets.

Emphasis on a high-technology path, however, need not mean total reliance on transferred technology or a wastefully extravagant domestic research and development system. A strong inclination to utilize both indigenous and exogenous technology sources often obscures the dynamic interaction that can be attained between the two when local innovation is tied to some importation of advanced technology. A sheer reliance on foreign technology without indigenous innovation is likely to result in economic dependence and stagnation, but the idea of applying a purely indigenous research and development without using foreign technology is a fantasy and a waste.

Since Korea seeks economic growth as one of its cardinal aims, it becomes clear why she chose the high-technology path. Some of the leading industrial sectors in a developing country are geared primarily to breaking the inertia of underdevelopment. They introduce technology quite advanced relative to their absorptive capacity, but feeder industries supporting the leading sectors do not necessarily require such advanced technology. Achieving a balance among all of these factors is a challenging task. At this point, therefore, it becomes desirable to look at the development of Korea's industry and her scientific and technological infrastructure in more detail.

INDUSTRIAL DEVELOPMENT AND USE
OF TECHNOLOGY IN KOREA

Korean industrialization began, for all practical purposes, in 1962 when the first Five-Year Economic Development Plan was launched. A few statistics can give some idea of the initial conditions in the period from 1959 to 1961: first, the GNP growth rate averaged 3.3 percent per year while population growth was 2.9 percent, leaving a 0.4 percent net increase in growth. The per capita GNP was under $100, so that altogether Korea offered a perfect example of an underdeveloped economy with accompanying stagnation. Second, a severe dichotomy existed within the economic structure: 65 percent of the total employed were in the primary sector, while secondary industries (mining and manufacturing) employed only 6.9 percent of the total. In terms of per capita output, primary industries represented 0.65 and secondary industries 2.60, when we use an index of one for all industry. This means that a mere 7 percent of the total employed population were engaged in relatively productive sectors. Third, the domestic savings-to-GNP ratio was only 3 percent; perhaps because of that, the Korean economy was, from the beginning, an open one vis-à-vis foreign countries for obtaining the necessary investments that accompanied the rise in the import of foreign goods and services and also, to some extent, the rise in exports to pay for the imports.

The Korean economy has, however, emerged after sixteen years with a completely new configuration. The GNP went up at an approximate rate of 10 percent per year to reach a per capita output of $1000, and the domestic savings-to-GNP ratio improved to over 20 percent. In 1977, exports, which were 90 percent manufactured goods, reached $10 billion, compared with less than $55 million in 1962. The structure of the economy also changed notably, in that agriculture's share in the GNP, which had been 44 percent in 1961, declined to 20 percent in 1976 while the manufacturing sector's share rose from 11 percent to 35 percent in the same period.

By the time the Fourth Five-Year Plan ends in 1981, it is envisaged with optimism that the GNP will reach $58.7 billion and a per capita output of $1512, making a twofold increase in net terms, as compared to that of 1975. The secondary and tertiary sector shares of the GNP will rise to 40 percent, leaving the primary sector with an 18 percent share. The export target for 1981 is set at $14.2 billion in 1975 constant prices, representing an annual rate of growth of 16 percent in net terms.

Initially, Korea's industrialization strategy emphasized the development of light, labor-intensive industries to absorb the labor force from the primary sector. The effective demand in the primary sector for industrial products was, however, all too slight, so it was necessary to be outward looking in terms of capital, market, and technology. Korea, therefore, did not choose the import-substitution-followed-by-exports type of industrialization for development, but instead the two were undertaken almost simultaneously, particularly when the first long-range economic development plan went into effect. The apparent success of this bold approach can be attributed to several factors:

1. The amenability to training or the absorptive capacity of the labor force in dealing with technologies that were relatively sophisticated (even during the Korean War over 15 percent of the total population was involved in acquiring formal education, as compared to only 6 percent before the liberation from Japan in 1945)
2. Close trade relations with the United States and Japan, both big markets
3. Full exploitation of the technical advantage of being latecomers in industrialization
4. A capacity to adapt to the international economic environment that was actively supported by the government via the creation of a favorable investment climate for foreigners

The most conspicuous constraints on the scheme for rapid industrialization progress were the deficiencies in the social overhead sectors: the infrastructures for industrial development were very poor, so this was the area on which the government placed greatest emphasis for quick and decisive action to build up roads, ports, communications, and other essentials for development. The government also worked on expanding educational facilities, particularly for technical education. About 50 percent of the total induced foreign capital was spent in this area as well as over 70 percent of the total public loan funds from overseas.

During the first economic development plan (1962–66) there were many occasions when considerable debate—all justifiable—occurred in the course of choosing a proper technology, but the choice of technology necessarily had much to do with the scale of a project. For instance, the first integrated iron and steel plant with an annual capacity of about 400,000 tons per year was contemplated at the time the first plan was formulated. Something less than the optimum size was considered; at that capacity it was impossible to introduce any modern production systems, so that from a techno-economic point of view, it was virtually impossible to set up an efficient integrated plant. Thus, the establishment of the plant had to be deferred until the second five-year plan when it would be justifiable to establish a plant with a capacity sufficient to include, for example, a continuous tandem rolling mill. It was necessary to wait another several years before a highly sophisticated continuous casting mill was added to the steel-making plant. This addition was introduced only after sufficient experience in operation and maintenance of the conventional methods was acquired.

The Second Five-Year Economic Development Plan (1967–71) placed emphasis on initiating the lead sectors approach and pushed forward with the development of the basic chemical industries such as fertilizers, cement, and petrochemicals as well as the iron and steel industry. What was attempted was the initiation of a growth momentum through these sectors so that a dynamism could make itself felt within Korean industry. These industries by their very nature are highly capital-intensive and need huge infrastructures that have to be supported by the government because they are essential to the foundation upon which the high-linkage industries can be built. In formulating the Second Five-Year Plan, we introduced a series of bold quantitative tools for formulating development models that were of the greatest importance in articulating our socioeconomic goals. These tools enabled us to develop the growth path we would travel to reach our goals, to identify the major constraints, and to formulate investment programs. They included, among others, an input-output model covering forty-three domestic production sectors and thirty-four import sectors, plus four value-added and seven final-demand columns, a medium-term macroeconomic model, as well as a short-term stabilization model. For several new but key industrial sectors such as steel and petrochemicals, what is known as a mixed-integer programming model was applied.

This approach, together with the use of foreign experts, was intended to allow greater latitude for debate on the plan so that a defensible strategy would of necessity follow. Noteworthy is the fact that the experts often turned out to be too conservative at both the micro- and macro-planning levels. The dynamism of a developing country when it reaches its momentum is difficult for foreign experts to accept and defies statistical prognostication.

One pressing problem in developing these lead sector industries was the question of whether or not they could be operated at full or at least near full capacity. It was found that the allowance was extremely small because of the fact that the cost of the capital for these industries, which mostly originated from abroad, was very much higher than for the advanced countries. Recognition of this hard fact of life had much to do with the makeup of any industrial project. Becoming overly optimistic can lead to the acquisition of a burden of nonproductive capital tied up in excessive capacity, which can slow the forward momentum.

The Third Five-Year Economic Development Plan (1972—76) followed later in more or less the same direction of industrialization with greater concentration on heavy and chemical industries of a capacity to enjoy an economy of scale. Agriculture and social services also received some attention. This orientation necessitated the introduction to industry of a greater number of new and higher-level technologies on an order of magnitude never before experienced. It was an irreversible decision so far as science and technology development was concerned. It was an issue of survival or extinction in ever-stiffening international competition.

Korea's experience in the past decade, with particular reference to the relation between commodity exports and royalty payments for induced foreign technologies, leads us to believe that an adequate supply of appropriate technologies, often advanced because of our development stage, is the essential factor in enabling industry to produce the goods and services that can gain better access into international markets.

For this reason, one of the two pillars of Korea's science and technology development policy hinges on attaining a capacity for the proper selection, digestion, and adaptation of imported technologies. Throughout the previous three five-year development periods, foreign technologies were often mingled with the inflow of capital. The situation has changed, however, and the explicit need for technologies, as clearly distinguished from capital, has been recognized.

A plan for development like that envisaged for Korea requires science and technology institutions to implement goals. The Korean approach to institutional frameworks was somewhat daring. It included, among other things, the establishment of: (1) the Ministry of Science and Technology (MOST) in 1967 as the central planning, coordinating, and promotional body in the government; (2) the Korea Institute of Science and Technology (KIST) by a special law in 1966 as an autonomous multidisciplinary industrial research institute chartered as a contract research organization; (3) the Korea Advanced Institute of Science (KAIS) in 1971, supplementing the many existing universities and colleges, as a mission-oriented postgraduate school in selected applied scientific and engineering fields under the jurisdiction of the Ministry of Science and Technology instead of the Ministry of Education; and (4) a huge number of vocational training institutes along with technical high schools to meet the rapidly rising, almost explosive demand for skilled workers and technicians. Let us consider the first three institutions in more detail.

The emergence of MOST spearheaded the enactment of several very important laws for the growth of science and technology for development. They include first, the Science and Technology Advancement Law of 1967, which defines the basic commitment of the government to support science and technology and to provide policy leadership. Second, the Law for the Promotion of Technology Department of 1972 provides fiscal and financial incentives to private industries for technology development. Third, the Engineering Services Promotion Law of 1973 promotes local engineering firms by assuring markets on the one hand and performance standards on the other. Fourth, the National Technical Qualification Law of 1973, through a system of examinations and certifications, sets standards for those who practice technical and professional skills. Fifth, the Assistance Law for Specially Designated Research Organizations of 1973 provides incentives in legal, financial, and fiscal terms for research institutes in specialized fields where the government and private industry place particular emphasis, such as shipbuilding, electronics, communications, mechanical, and materials engineering. And sixth, the Law for the Korea Science and Engineering Foundation of 1976 gives a legal basis for the establishment of a foundation to act as prime agent for strengthening research in basic and applied sciences and to facilitate more rapid application of science and engineering to national needs.

The Korea Institute of Science and Technology was brought into being to bolster the industrial sector in those areas where the national economic development plan placed its emphasis in order to eliminate the bottlenecks hindering further growth. Through special legislation KIST was made a contract research organization so that marketing principles would prevail in the realm of research and development. Thus, researchers became problem-oriented and the underwriters of R&D came to realize the importance of using R&D results.

Before taking on any research, KIST carries out a comprehensive study to ascertain what needs exist. In formulating the major thrusts of the KIST program in the very beginning, a major study of this nature was conducted on 600 industrial plants and related organizations, covering twenty-five industrial sectors. It took eight months to complete and involved eighty specialists, including twenty-three from abroad. The survey helped identify the main concentration areas for the institute's initial period of operation. The areas included material and metallurgical engineering, food technology, chemistry and chemical engineering, electronics, mechanical engineering, and industrial economics and management.

Based on this selection, such important decisions as those concerning staffing, equipment, and facilities were made to ensure that the institute would be properly equipped to solve both the country's current and emerging problems. Such studies are carried out periodically to cope with industry's ever-changing needs and to keep KIST a dynamic partner in development. Among the systems set up within the institute in the attempt to make it a viable institution serving the ends for which it was designed were a project development program to sell the scientific research concept to industry and to help industry formulate the questions to ask the institute, a cost accounting system to gauge input as compared with performance, a multidisciplinary approach to avoid the rigidity inherent in departmentalization, and a large endowment fund so that the institute could undertake long-term research for which no particular client could be found.

As industry grew, however, its technological requirements increased in level and diversity, and laboratories such as those in shipbuilding and petrochemicals, which existed as integral parts of KIST, were no longer able to render the necessary technical support to industry. The rapid growth of these industries required an independent research organization specific to each industry and problem area. Creating such institutions from scratch would have been a formidable task, so the existing small laboratories at KIST were used by spinning them off the mother institute, thereby allowing them to inherit not only what had been accumulated, but also a working management system and philosophy which are all too often missing or in an amorphous state in a new organization.

Oftentimes a considerable inertia accompanies the implementation of a new program through an already functioning institution.

This perception on the part of the government led to the establishment of a completely new institution, the KAIS, when it became clear that new high-level scientific and technical manpower was needed for industrialization. Rather than depend on an existing institution of higher learning to train the kind of manpower industry required, KAIS was organized to provide graduate training in applied science and technology.

Since its first classes opened in September 1973, KAIS has produced over 500 master's degree graduates who have filled many important positions in industry, academia, government agencies, and research institutions. At the time it was established, everyone worried about what was to be done with so many graduates, but from the beginning the demand has exceeded the supply by far. Industrial demand for graduates has attained such proportions that KAIS graduates must be rationed.

The importance of industrial research has only recently been recognized, despite its obvious role in economic development; in Korea, industrial research is even more important in framing economic development strategies. It is essential to the realization of a nation's industrialization goals within the contexts of the global and the regional economy; if it is properly carried out, it can help in setting reasonable goals.

In striving toward these goals, Korea has made substantial achievements, through trial and error, in improving national scientific and technical capabilities, in bringing innovations to the administration and support systems, and in increasing as well as orienting R&D investment. The total science and technology effort was intended to effect a structural change in the economy from a simple labor-intensive one to a more viable technology-intensive structure and later to the development of a brain-intensive one. In other words, these efforts were directed toward accelerating the transition of science and technology's role from supporting national economic development to leading such development on a foundation of a technologically self-reliant economy.

A technological latecomer like Korea for the most part has no choice but to follow the technological lead. Once industrialization was the national goal, high technology was the inevitable course rather than the choice. It was a sort of "blind force."

In conclusion, let me make a few general observations. First, the notion that industrialization in a less-developed country does not create enough employment to make it worth trying has limited validity. In the case of Korea, industry has provided at least one-third of all jobs created since 1963. Second, the notion that less-developed countries do not require high technology if they set their targets more on agriculture than industry also has limited validity, particularly when there is little arable land to support a large population. Agriculture does require a considerable array of what might be called high technology, as is the case, for instance, with developing high-yield varieties that can suit particular ecological and environmental conditions. Third, the notion that less-developed countries do not require domestic R&D, but rather injections of technology from developed countries also has limited validity as domestic R&D is a prerequisite in enhancing the technological literacy necessary to make it possible to take advantage of foreign technologies.

In a developing country with high selectivity in terms of sector, size, and degrees of capital and technological intensities, industrialization can bring about many essential improvements that are not likely to be achieved otherwise. The problems that need to be solved in a developing economy often require high technology to set development in motion as forcefully as possible and thus overcome apparently insurmountable obstacles.

Chapter Seven

The National Development of Ghana

Robert Dodoo, Jr.

Ghana emerged out of the colonial era with its attendant colonial economic development and structure into nationhood in 1957. Ever since independence, one of the prime preoccupations of decisionmakers and planners has been the search for solutions to the problems of national socioeconomic development. Since 1957, therefore, the country has not lacked short- and long-term development plans and strategies, aimed at improving and diversifying agriculture, developing infrastructure, expanding industry and accelerating the overall pace of growth in the other sectors of the economy. These measures have all been targeted at achieving the ultimate national goal: economic self-reliance and self-sufficiency and the raising of the standard of living of the people as a whole.[1]

Twenty years after political independence there has been limited success, in the face of seriously recognized constraints, toward the realization of this ultimate goal. Some of the constraints definitely could be attributed to the pattern of preindependence economic structure and activities. But, to a large extent, most of the post-independence development ills have been the result of frequent changes in political direction and the mismanagement of the economy by successive decisionmakers. The key to the solutions to the development problems, and its accompanying relevant measures and mechanisms for reaching the goal development, seems to have eluded decisionmakers and economic planners.

The key to the success of development plans, in particular the Five-Year Development Plan (1975/76–1979/80) in Ghana, and to national prosperity, apart from the spirit of self-reliance of the people, may perhaps lie in the effective mix of three factors—technology including institutional linkages, natural resources, and capital. But of the three, technology in the developing economy of Ghana seems to be the key element, since the adoption of new and appropriate scientific and technological techniques can in many instances make up for a deficiency in resources and substantially reduce demands on scarce capital.

A 1973 study conducted jointly by Ghana's Council for Scientific and Industrial Research, the universities of Ghana, and the U.S. National Academy of Sciences recognized that science-based technology is directly linked to the development of a nation. The participants outlined their rationale as follows: technology comes from discoveries of science, and growth of technology helps to improve the quality of life for the people of a country.[2] They pointed out, however, that even though a relation certainly exists between technology and the quality of life, it cannot be concluded that development will automatically parallel the growth of science. The functional relation, they emphasized, depends on what kinds of science grow, on the incentives offered, and on the machinery available for feeding new knowledge into a complex interlocking system for improving the health of people, the comfort of their life, and other things that add up to their happiness.

Lately, and in the course of the search for solutions to economic development problems, it does seem that a general awareness of the virtues of scientific research and technology and their application to development needs is finally emerging in Ghana. The present government, in particular, has finally noted that inadequate and ineffective technological inputs constitute one of the major constraints in the country's development. This recognition stems from the realization that the technology being employed in many processes in the country, especially by the average farmer and the small processor at the village level within the rural areas or at the fringes of the urban centers, is usually obsolete and has resulted in low productivity per capita.

BACKGROUND TO ECONOMIC DEVELOPMENT

Basic Characteristics of the Economy

The basic character of the Ghanaian economy poses tremendous technological problems and opportunities. The economy remains essentially agricultural and rural. The dominance of agriculture, which in the Ghanaian context is broadly defined to include livestock, fisheries, and forestry, is demonstrated by the following statistics: Agriculture accounts for more than 40 percent of the gross domestic product, employs about 60 percent of the working population, and provides some 70 percent of the total export earnings.[3] Another significant hallmark of agriculture is that cocoa, as a cash crop, singularly dominates all the other agricultural crops, thus critically influencing employment, foreign exchange, and government revenues.

Food production also poses another perplexing dilemma in the economy. Until the 1970s, when commercial farming was introduced, it was practiced only on a small scale and left in the hands of peasant farmers, who use few inputs besides land and labor. Peasant farming has remained essentially traditional, that is, based on shifting cultivation, at the subsistence level, and characterized by low income and low productivity. It has been estimated that this small-scale peasant system produces about 90 percent of the country's crop harvest, out of which only about 50 percent actually enters the commercial distribution system. The infusion and impact of improved practices, credit facilities, appropriate technology, and purchased inputs in food production has been very limited.

Another problem of the economy is the rate of population growth. Ghana's population has been estimated at nearly 10 million in 1975, but the rate of growth has hovered round 2.7 percent per annum. This rate of population increase coupled with the slow rate of growth in economic output has resulted in a stagnant economy as reflected by the per capita income. The labor force declined from about 41 percent of the total population in 1960 to slightly less than 39 percent in 1970.[4] This reduction in the labor force was due, especially among the youth, partly to the promotion of economic activities using imported technology to produce goods and services that compete directly with those produced in the traditional sector. In addition, there is seasonal unemployment in the traditional economic sector, especially in agriculture, where production is limited to specific seasons during the year. The reduction in the labor force thus constitutes an intolerable problem for the economic development and political stability in the country.

The absence of appropriate technological linkages between industry, agriculture, and the other sectors of the economy has contributed to the balance-of-payment difficulties facing the nation. This weakness has frustrated several attempts at reviving and stimulating the Ghanaian economy. Increases in consumption in one sector not only fail to generate economic activity in other sectors of the economy but rather generate increases in the marginal propensity to import.

There has also been a pattern of increasing inability to generate jobs due to the failure of the economy to achieve sustainable growth during the last decade. The rate of investment has been low and this, together with the slow technical progress, has had an adverse effect on employment opportunities. Urban unemployment in 1975, for instance, accounted for 10 to 13 percent of the labor force even though there has been a tendency for labor shortages to develop in the agricultural sector.[5]

In the sphere of industry the position of the government has been to encourage private enterprise, both foreign and Ghanaian. Foreign investors are encouraged to enter the modern industrial sectors and are granted incentives and guarantees under a Capital Investment Act and other official regulations.[6] The intention of the government, however, is to prevent foreign control and domination of industrial activities. Over the years, then, joint ventures and purely Ghanaian-owned enterprises have grown at the expense of the completely foreign-owned activities. These Ghanaian enterprises, unfortunately, have been marked by a shortage of efficient managerial and supervisory personnel essential to the proper utilization of scarce capital equipment.

Manufacturing has also changed from an activity based on the dominance of wood products and basic import substitution to one of light industries producing largely consumer goods mostly for the urban market. But the external orientation of the manufacturing sector has been significant in that most of the inputs and equipment have to be imported. As a result there has been an unholy alliance between the balance of payments and the level of capacity utilization.

With respect to the external orientation of the Ghanaian economy, the lack of foreign exchange constitutes a significant and most pervading constraint. Any significant growth in output has to be reflected in an increase in imports. Ghana has continued to support growth in gross domestic product by the importation of investment goods, intermediate goods, and some consumer goods. Agriculture presents a frustrating example of the external orientation of the economy. In a situation where agriculture plays a dominant role, it is a major weakness when there are no domestic facilities for supplying the vital supporting inputs such as fertilizers, insecticides, and equipment. A critic has noted that "an increase in the marginal propensity to import is one of the signs of technological backwardness in any economy."[7] To put it mildly it is, in the Ghanaian context, an expression of extreme technological needs.

Some Identified Constraints

The Five-Year Development Plan itself identified a number of other deficiencies that, in my view, require technological applications.

The infrastructure for delivery of essential inputs to all sectors of the Ghanaian economy is weak, and inappropriate, and therefore ineffective in stimulating and increasing the small entrepreneur's productive capacity. It has been noted that part of the problem arises as the result of inefficient methods of importation, slow clearing of goods, and inadequate transportation facilities essential for the speedy delivery of either imported or locally produced inputs. These problems are further compounded by poor operational planning and lack of effective coordination at various levels.

Further, there is a marked lack of access to institutional credit in the rural areas for the benefit of the small operators engaged in agriculture, industry, and other businesses. The financial institutions with their formal requirements, time-consuming procedures, and complicated administrative machinery have frustrated the small operators. The institutions have not been innovative in dealing with the small operators and have geared their activities toward the large firms within industry and commerce. It is therefore most difficult for the small operators, especially in the rural areas, to contribute significantly to the economic development efforts. However, a new move by the government to establish banks in the rural areas to cater specifically to the needs of the rural population may promote such contributions.

An assessment of the Ghanaian pricing mechanism urgently needs to be undertaken if the heretofore unsuccessful pricing system is to have the desired impact on productivity. The prices of commodities, when properly managed, influence the individuals in their attempts to reap the maximum benefit from their endeavors. Despite a governmental pricing policy, an increase in production has yet to be realized, especially within the agricultural sector.

The age-old problem of land tenure presents yet another obstacle to raising productivity. The existing land tenure system impedes land acquisition, particularly for large-scale farming. The multiplicity of claims on ownership and the insecurity of tenure restrict large-scale cultivation of the land and investment in land improvement.

The mixed economic structure of the country raises perplexing issues for the growth of productivity. Composed of both private and public sectors, the economy has been characterized, since independence, by a pattern of informal and unregulated economic activities and investments. The state participates in a number of ventures and controls some areas. Still other segments are entirely in private hands. Foreign control of much of the economy is pervasive.

One of the weaknesses of this unregulated economic system has been the implicitly strong open-market elements within the two sectors and the failure to inject appropriately recognized technologies, techniques, and measures to meet the requirements of the open-market elements and the demands of the centrally planned elements. Rather than employing science and technology to develop the nation's natural resources, Ghanaian entrepreneurs and managers in both the public and private sectors of the economy seem instead to depend upon the government to use the scarce resources of the state in providing most of the critical industrial inputs—finance, raw materials, and capital goods—necessary for ensuring growth and development in the various industrial activities in the country. Consequently there has been excessive liberalization as regards access to and utilization of foreign production processes and related resources in solving the problems of the production system.

Finally, the marked absence of any meaningful statistical data base for use in planning, plan-evaluation, and decisionmaking has been one of the major lacunae in the tools for decisionmaking. In some cases this weakness contributes to the failure to establish clear goals and guidelines for the efficient performance of achieving the overall targets that have been assigned to institutions.

DEVELOPMENT GOALS AND OBJECTIVES

The government is committed to the concept of economic self-reliance. It is, therefore, the intention of the government, during the current Five-Year Development Plan (1975/76−1979/80) effectively to implement that concept by a structural transformation of the national economy to achieve three basic goals: (1) capturing command of the heights of the economy and placing them firmly in the hands of Ghanaians through the deployment of state power, (2) promoting a stable economic growth and development, and (3) ensuring that the fruits of development are equitably distributed to improve the quality of life for all Ghanaians.[8]

In Ghana, as in other developing countries, one of the primary concerns has been the need for rapid expansion of the output per capita. Therefore, one of the declared objectives of the government is to ensure that output grows faster than population. This requirement is a prerequisite for the realization of economic and other fundamental goals since it makes possible improvement in the standard of living of the present and future generations of Ghanaians, especially for those subgroups within the lowest incomes and in the rural areas. This requirement acts to reconcile competing demands for higher wages—the result of inflation push—with some improvement in the standard of living.

Ghana generally is characterized by very low per capita income and insufficient total growth, and the overwhelming majority of the people live in miserable conditions. It is considered imperative, then, to attain a fair and equitable distribution of income and wealth, through the provision of gainful employment for all Ghanaians who are willing and able to work. This objective is of immense importance; to avoid social unrest and political instability, all the available resources of the nation must be effectively employed to reduce the high unemployment level and excess capacity in industries, and to achieve increasingly productive use of our land, water, and other natural resources.

The greatest task that faces any country like Ghana, which has attained political independence after many years of colonial rule, is the quest for economic independence, because without economic independence, political independence becomes limited in scope. The essential features of the quest for such a goal within the Ghanaian context are: [9]

1. The creation of effective links between the sectors of the economy so that development becomes mutually reinforcing
2. The capturing of the commanding heights of the economy by Ghanaians and ultimate independence from foreign aid
3. The achievement of economic independence in food and other essential consumer and investment goods in order to reduce undue reliance on imports and increase dependence on local substitutes
4. The provision of local substitutes for any goods that are currently being imported and whose consumption can be postponed without severe damage to life or to the country's own ability to produce
5. The diversification of the commodity base in the export sector in order to seek new markets

To accomplish the objectives set forth by the government, detailed development programs or projects along with targets have been carefully identified in the Five-Year Development Plan documents. These programs are expected to be implemented by various government departments, public corporations, and quasi-governmental institutions in the country. The plan has been laboriously drawn up, but there has not been an accompanying technology plan to give effective support to the projects envisaged and the targets set. This weakness ought to be corrected. A major step would be to analyze the current development programs and other statements made on national goals, to compare implied needs with projected activities for science and technology, and to determine the science and technology requirements of each. It would then also be necessary to assess and delineate the scientific and technological capabilities and institutions demanding priority attention during the current and subsequent development plan periods.

NATIONAL SCIENTIFIC AND TECHNOLOGICAL INFRASTRUCTURE

The delineation of technological requirements within the developmental model is only a first step toward achieving growth and development in Ghana. Technology requires a viable economic and social structure, efficient supporting services, and a strong infrastructure for it to have the desired impact on national development. It is the existence of a sound scientific and technological policy, research and development establishments and capabilities, educational system, training capabilities and management potentialities that count most. And it is precisely in these areas that, in Ghana, the technological needs are most apparent.

Science and Technology Policy

There has been, as far as can be judged, no conscious attempt by decisionmakers, economic planners, and administrators to articulate a science and technology policy in any of the past development plans. Some believe that although a separate chapter on science and technology was not included in the Two-Year Development Plan (1972/73—1973/74) the government was still fully aware of the key role that scientific research and its applications would play in the fields of agriculture and industry for the success of the plan. Furthermore, it is claimed that the existence of research institutes that are steadily expanding and that are expected to gear their programs to the economy ensured that the absence of a specific chapter in the Two-Year Development Plan did not in fact constitute an omission.[10] I disagree on this issue.

When the Five-Year Development Plan (1975/76–1979/80) was drafted, its authors recognized the need to stress the direction of science and technology during the plan period. Thus they underscored the fact that the failure to accord science and technology policy a place in previous development plans constitutes not just an omission but also indicates a failure to recognize the role that science and technology should play in development.

Hence, for the first time in the history of planning in Ghana, the Five-Year Development Plan restored the full rights and responsibilities of science and technology. In so doing it recognized the limitations of previous scientific and technological efforts by asserting that research has been rudimentary, isolated, and barely related to pressing national needs so far as most of the farmers, fishermen, and consumers are concerned. Where some spectacular and relevant results have been obtained from research institutions and facilities, these have hardly been properly disseminated or translated into the types of projects that would have a significant impact on national economic growth and development. There are no effective channels for communicating the problems of farmers and consumers through a coordinating body (which should establish national priorities, publicize, and allocate the problems and tasks) to the agencies or institutions that possess facilities and capacities for undertaking the required research. These problems are compounded further by the absence of an effective extension service to reach the farmers and fishermen.

The science policy contained in the development plan is, however, only a mere outline and not a dynamic policy. A dynamic national policy for science and technology must include a technology plan; it must be designed to use the resources of the nation for orderly growth and change. Such a concept could be accepted in Ghana, but, given the conditions in the nation, it may generate fear that abrupt change may waste what has already been achieved and may severely disturb the lives of the people most directly affected by the growth process. Though Ghana may not yet be receptive to such a dynamic policy and its implications, it is still important to aim at it.

Research and Development
Institutional Capacity

Research and related activities have been institutionalized in Ghana since 1958, the year in which the Research Act was promulgated and the National Research Council established. In 1963, this body gave way to the Ghana Academy of Sciences, the predecessor of the present Council for Scientific and Industrial Research (CSIR).

The CSIR is the nation's central research coordinating organization. It is a governmental body reporting directly to the Ministry of Economic Planning and charged with the responsibility for: [11]

1. Advising the government on scientific and technological matters
2. Organizing, co-ordinating, and encouraging research
3. Initiating and executing, through its research institutes, projects that would have direct impact on national development

The following are research institutes or units or projects that, under the CSIR, undertake applied research: Animal Research Institute, Building and Roads Research Institute, Crops Research Institute, Food Research Institute, Forest Products Research Institute, Soil Research Institute, Water Resources Research Unit, National Atlas Project, Herbs of Ghana Project, and Northeast Ghana Savannah Research Project.

In addition to the work done by the CSIR, applied and basic research and related activities are carried out by some of the faculties and departments of the nation's three universities, as well as various departments of government ministries, such as the Cocoa Research Institute of the Ministry of Cocoa Affairs, Fisheries Research Unit, Animal Husbandry Division, Veterinary Services Division, all of the Department of Agriculture, and the Geological Survey Department, under the Ministry of Lands and Mineral Resources.

A few government departments such as the Marine Fisheries Research Unit of the Ministry of Agriculture and the Volta Lake Research and Development Project, both of which were established jointly by the Ghana government in collaboration with U.N. bodies, are full-time specialized research units or projects, conducting investigative research on problems related to their normal operations.

Ghana's institutional framework and capacity for scientific and technological research appear adequate at first glance. It has been observed, however, that most of the research programs being undertaken lack a coherent framework matched to the objectives and time scales of development plans. Some projects are highly academic in their approach, while others give rise to fragmentation and duplication of effort, and are essentially open-ended surveys or data collections in nature. [12]

Domestic research has had only limited effectiveness in producing results that contribute directly to national development. The linkage between the research community and the end users is limited, and consequently the end users have not acquired the habit of making demands on the research system. There may be some justification in attributing the limited impact of the research and development establishment to insufficient research personnel, shortage of equipment, and inadequate domestic funds and foreign exchange to carry through research activities successfully. But government action contributes some problems. The exigencies of the economic situation in the country prevent rational allocation of funds to research and development. The government, further, does not provide the CSIR with independent research funds. Instead, it provides funds to the various research institutes independent of the CSIR. The funds in most cases are allocated for supporting ongoing research programs regardless of their relevance; hence government action does not enhance the advisory, research, and coordinating role of the CSIR.

The CSIR requires revitalization, which requires governmental action to centralize its research and development funds within the CSIR so it can:

1. Coordinate research in all its aspects in the country
2. Strengthen its planning and analysis of research capabilities
3. Promote the dissemination of research results to the productive sector
4. Foster innovative research of relevance to the immediate requirements of the productive sector
5. Guarantee the training of research personnel
6. Ensure the availability of equipment needed for research

Technology Transfer
The provisions of the Investment Policy Decree of 1975 indicate that the government intends to use the full resources of the state to foster a mixed pattern of economic development and to encourage and ensure maximum development, including the choice and acquisition of technology, within the following five economic levels and activities:

1. Exclusive control by the state of public utility and infrastructural projects and the manufacturing of arms and munitions
2. Joint state/foreign ownership in timber and mineral extraction projects as well as banking and insurance
3. Joint Ghanaian/foreign ownership involving various commercial, industrial, and agricultural projects (e.g., shipping, manufacture of furniture, petroleum distribution, fish and shrimp trawling)
4. Full Ghanaian ownership of enterprises that do not require large amounts of capital or involve relatively complicated technology,

such as retail trade, business representation, baking, charcoal manufacture, and tailoring of garments
5. Enterprises earmarked for complete foreign ownership which involve complicated technology, require foreign exchange, and are export-oriented, like the huge Volta Aluminium Company (VALCO)

Technology from abroad enters Ghana mainly through the first four levels of economic activity. A further source of technology for Ghana is technical assistance programs sponsored by international organizations such as the United Nations or the USAID and the training of Ghanaian students in foreign countries.

The basic technology transfer problem in Ghana, and a critical constraint to development in general, is the absence of a concise and comprehensive long-term policy on technology development and transfer supported by an institutional framework or mechanism for comparing and evaluating foreign technologies to meet long-term goals.

The Council for Scientific and Industrial Research is not involved in investment decisions. That role appears to be included in the functions of the Capital Investment Board (CIB) and the Ghanaian Enterprises Development Commission (EDC) whose day-to-day operations involve decisionmaking relating to the establishment of various industries and business enterprises. The legal and financial implications of industrial projects and business activities, in both the public and private sectors, do get scrutinized by government legal experts and other financial personnel at the CIB, the EDC, and other cooperating financial institutions. However, there is very limited and critical technological evaluation accompanying the legal and financial considerations.

A number of other considerations have further hindered the development and transfer of technology in Ghana. Prominent among them is the lack of foreign exchange or domestic savings needed to import technology. Most technologies in the industrialized countries are owned by private companies on a patent basis. These companies would normally release these technologies under licensing arrangements or would sell the technology in package form. Invariably, this method involves additional financial burden on the recipient country. Second, Ghana may need financial aid from an industrialized country in order to finance importation of technology. Such financial aid often does not leave the recipient free to choose the most appropriate technology. Rather, conditions are imposed on the recipient country to order all the equipment from the donor country.

This limits the scope of the recipient country in choosing an appropriate technology. Third, because of the restrictive practices of foreign companies (mostly multinational corporations), technical knowledge cannot easily be disseminated to other would-be users within Ghana. This inhibits the spread of technical knowledge and the replication of imported technology, and also creates a costly need to import the same or similar technology over and over again. Fourth, a lack of intermediate-level skilled personnel makes it difficult to take advantage of technology transfer.

In view of the preceding conditions there is the need in Ghana to develop the means for strengthening the scientific, technological, and administrative infrastructure to generate and sustain indigenous science and technology capabilities as well as the transfer of technology itself. This would require the design of a clear policy that would encompass external technological inputs on the one hand and accelerate indigenous technological activities on the other.[13] The overall objectives of the policy should include:

1. Implementing a policy of technological competition by maximum diversification of supplies
2. Laying down basic principles for the fixing of standards and conditions for contracts involving transfer of technology
3. Effecting a program of technical cooperation between Ghana and other Third World countries
4. Channeling new technology in priority sectors of the economy
5. Analyzing the functions of institutions in Ghana that engage in technology evaluation, upgrading the quality of their work, and enlarging their technological capabilities

Appropriate Technology

In the long run the transfer of technology can serve very limited development purposes in view of Ghana's foreign exchange difficulties and the restrictions inherent in the technology transfer system. For the immediate and future requirements of the nation, it would be desirable to develop technology appropriate to the needs and conditions of Ghana, that is, labor-intensive technologies or devices, simple but efficient technologies requiring minimal initial capital investment but producing high returns, technology that relies heavily on local inputs both in manpower and materials, and technology that would help in extracting the most out of available energy resources.

To meet the preceding technological requirements would require: (1) an assessment of the primary needs of the local population; (2) the demands for products and services to meet these needs; and (3) a subsequent analysis of the technologies that could economically cater to these demands, making optimum use of local resources.

Unlike many discussions of appropriate technology, these preliminary considerations have the virtue of covering both the demand for and the supply of technologies. Such a preliminary approach is essential because, in the developed market economies, the division of labor and the interconnections between the public sector (including the universities) and the private sector are such that the results of basic and applied research are readily translated into the production of marketable products, whereas in Ghana this interconnecting link is missing and the gap must be bridged.

Some relatively minor but potentially useful evolution of traditional technology activities has been undertaken by various pre-university educational institutions and by individuals working on their own farms and gardens. These traditional technologies are oriented toward solution of local problems of production activities in the rural areas (for example, farm implement manufacture, food grain storage, local crafts, informal construction work). It is generally agreed that there is considerable potential capability for upgrading the existing traditional technologies. The Technology Consultancy Centre (TCC) of the University of Science and Technology is gearing some of its efforts toward this end.

The reference to the optimal use of the local resources is not to be misconstrued as meaning reliance entirely on indigenous technologies. The really important objective is to make the maximum use of technologies appropriate to Ghanaian conditions.

Manpower Requirement

There is a growing demand in Ghana for capable and competent managers who can handle efficiently the increasingly complex problems of modern industry and associated sophisticated equipment. The lack of trained managers results in young graduates from the universities being placed directly in managerial positions. They often do not measure up to the responsibilities thrust upon them.

Furthermore, as a result of the Government's National Investment decree, all small-scale industries are reserved for Ghanaians only. The success of these industries and thereby their contribution to development fall on Ghanaians. However, most of the new managers and entrepreneurs have yet to develop fully the initiative, resourcefulness, and managerial qualities required to make a success of the businesses.

The universities of Ghana are a major source of science graduates and of managerial and technical personnel in the country. The output of science graduates appears adequate; however, most have acquired a basic science education and know little applied science. Thus, although the number of qualified research scientists in the country has steadily increased over the last decade, there are still critical areas where the number of specialists is small. Moreover, there has been no organized attempt to identify the critical areas of manpower shortages and the applied sciences and disciplines essential for supporting the national development programs.

The government of Ghana is very much concerned that research undertaken during the Five-Year Development Plan period be made relevant to the objectives of the plan. Therefore, research and development institutions clearly must master the art of defining their objectives, programing, budgeting of individual research programs, and project evaluation and accountability. It seems imperative for the institutions of higher learning, including the research bodies in the country, to train managers for research projects and the transfer of research results to the ultimate users.

In the past there has been the tendency to underestimate the need to train factory workers to undertake lower level management decisions and responsibilities; thus, there is a persistently weak link in the chain of effective management and the implementation of management decisions. Such a fault must be remedied.

In general the following are needed: (1) practical/factory attachment training programs in the formal educational system, (2) training of managers and engineers already employed, and training the critical class of middle and top grade personnel, and (3) training in research management.

CONCLUSION

Ghana inherited some of the scientific and technological gaps in the development process from its colonial past. These gaps, however, have been allowed to persist and have even been widened by post-independence economic, social, and political constraints. The result has been an inevitable retardation of the overall development process in the country.

Modest efforts have been made since 1957 to develop scientific and technological infrastructure in Ghana. But the momentum and the awareness to develop and apply science and technology toward solution of development problems and in furtherance of development goals and objectives have been great only during the past decade.

The philosophical principle of national economic self-reliance enunciated by the National Redemption Council in 1972 when it assumed the reins of power has permeated all aspects of the measures and programs intended for achieving growth and development in the economic and social areas. In the current national development model a bold effort has been made to define clearly the present and, to some extent, the future directions of growth and development. Steps have also been taken to delineate some of the constraints inhibiting development in the country, thus, in effect, indicating areas of technological needs.

Appropriate technologies and techniques seem to hold the key to national development aspirations. However, a number of technological constraints—weaknesses and needs—must be carefully delineated for the desired national development goals and objectives to materialize. But the identification of technological constraints and requirements alone is by no means a guarantee for achieving growth and development. It is only when it is coupled with sound science, technology, and management policies for generating and ensuring growth; the strengthening of our indigenous scientific and technological capabilities, educational system, training facilities, research and development institutions for new technology development, transfer and adoption of appropriate technologies; and the mobilization of human resources (scientific and technological manpower), that development becomes significant and self-sustaining.

The future then demands the formulation of a dynamic science and technology plan that would provide the technological support for directing short- and long-term development programs.

NOTES TO CHAPTER SEVEN

1. I.K. Acheampong, "Address on the Launching of the Five-Year Development Plan" (Accra, April 18, 1977).

2. Joint Council for Scientific and Industrial Research, Universities of Ghana and U.S. National Academy of Sciences, *Workshop on the Role of the CSIR in Determining Science Policy and Research Priorities in Ghana* (Accra, 1973), p. 1.

3. *Five-Year Development Plan—1975/76—1979/80* pt. 1 (Accra: Ghana Publishing Corporation, 1977), p. 1.

4. Ibid., p. 1.

5. Ibid., p. 3.

6. *Investment Policy Decree* NRCD 329 (Accra-Tema: State Publishing Corporation, 1975), pp. 3-6. Also Capital Investment Board, *Notes on Ghana's Investment Policy* (Accra-Tema: State Publishing Corporation, 1975), pp. 1-7.

7. M.N.B. Ayiku, *Industrial Research Opportunities during the Five-Year Development Plan* (Accra, 1977), p. 5.

8. *Five-Year Development Plan*, pt. 1, pp. 26-27.

9. Ibid., p. 29.

10. Council for Scientific and Industrial Research, *The Present Situation and Future Prospects for Science and Technology Policy in Ghana* (Accra, 1972), p. 11.

11. Council for Scientific and Industrial Research, NRCD No. 293 (Accra, 1968).

12. E.S. Ayensu et al., *Report of the Committee to Review the Research Function of the Council for Scientific and Industrial Research* (Accra, 1977).

13. Jean-Louis Schmidt, "The Transfer of Technology from Industrialized to Developing Countries," *Courier* (EE–ACP Publication) 39 (1976): 38–39.

✳   *Chapter Eight*

The Development Problem, Strategy, and Technology Choice: Sarvodaya and Socialist Approaches in India

V. V. Bhatt

Much before independence, there were two development ideologies in India—the Sarvodaya one (the word was derived from Ruskin's *Unto This Last*)[1] represented by Gandhi and the socialist one with Nehru as its spokesman. The major objective of both was to remove poverty and unemployment through socioeconomic development in an environment of freedom and democracy. However, there were major differences between the development strategies of each and their implicit technology policies.

For Gandhi, the immediate problem was poverty and it had to be tackled directly through the provision of full employment in traditional sectors—agriculture and cottage industries. The modern sector's development was to supplement and reinforce the development of the traditional sector. The emphasis, thus, was on upgrading traditional technology and on adapting modern technology in a manner consistent with the local situation and the development objectives. The major instruments for the purpose were education and scientific-technological research that was organically related to the identification and solution of concrete socioeconomic problems.

Nehru's strategy was large-scale industrialization with an emphasis on capital and heavy industry. Cottage industry was to be tolerated as a means of providing employment in the short run, but the main objective was to develop large-scale modern industries and techniques that would *supplant* the traditional sector. The dominant sector was to be the modern sector based on modern science and technology.

The views expressed in this chapter are solely those of the author and do not necessarily reflect the official opinions or views of the World Bank or its affiliates.

The major instrument visualized for development was scientific and technological research that gave mastery over modern technology. There was no emphasis on upgrading of traditional technology nor was there any appreciation of the problem of adapting modern technology to the Indian environment. Socioeconomic development was to be a planned endeavor with a major role assigned to the state; the objective was to evolve a socialistic pattern of society.

Since independence, the dominant ideology has been the socialist one, evolved by Nehru himself. However, the Sarvodaya ideology has been kept alive out of respect for Gandhi by institutions that were created by the state for the promotion of cottage industry and by the Gandhi followers. The wheel has come full circle in the sense that the present government—the Janata Government—has adopted Gandhi's strategy with emphasis on agriculture and cottage industry and small industry.[2] Yet the problem of poverty and unemployment still remains the major problem after more than twenty-five years of planned development.[3]

This essay seeks to trace the evolution and to discuss the elements of technology policy that are relevant for India and generally for the less-developed countries. Gandhi's approach and Nehru's socialist ideology are presented first. Socialist ideology as embodied in the Five-Year Plans is the subject of the third section. The fourth section draws attention to the fact that India still has not evolved a viable technology policy as an integral part of development strategy and discusses the nature of relevant technology policy for socioeconomic development. Finally, some case studies to illustrate the implications for technology choice of the two approaches are presented.

SARVODAYA PERSPECTIVE

Gandhi was painfully aware of the degrading, dehumanizing poverty of the Indian masses. His principal objective was to create socioeconomic conditions that would be consistent with the moral and mental development of each individual and family. He was in search of an "immediate, practicable and permanent" solution to the problem of poverty. This poverty and "enforced idleness" were related and hence *full* employment—work that gave a "living wage"—was to be the basis of his approach. It was not possible to provide "decent minimum" living conditions without work and employment, and even if it were possible, it would be degrading and demoralizing both for the individual and the society to have parasites—individuals eating without work.[4] "By the sweat of thy brow thou shalt eat thy bread." From this Biblical injunction, seen by Gandhi in the Gita, came Gandhi's concept of bread-labor.[5] For the full moral and mental development of the individual (the individual was, for Gandhi, "the supreme consideration") it was essential to find work that gave him or her opportunity for self-expression and development of creative intelligence; social relationships would be corrupted to the core if the individual obligation to work were not recognized as a social duty.[6]

Gandhi believed that unless the means of production for basic necessities were owned by families or a cooperative of families, no individual could be really free to mold his life. Hence, agriculture and cottage industry were to be owned and operated by families (without hired labor) or by cooperatives. The village had to be industrialized and resuscitated, for the village was to be the basis of a just, nonviolent social order, founded on willing cooperation and the elimination of social conflict. Whenever conflicts arose they were to be resolved by peaceful and nonviolent means such as civil disobedience and noncooperation.[7]

Gandhi recognized the need for some large industries. But then, they were to be organically related to the needs and requirements of agriculture and cottage industry.[8] Large industry was to be owned and operated by the state; where, as a matter of history, individuals owned such industries, they were to act as trustees. The doctrine of trusteeship was to be enforced by law and if need be by nonviolent noncooperation on the part of the workers.[9]

The major role of the state was to facilitate the evolution of such a nonviolent social order by performing such social tasks as were essential for coordinating and reinforcing the development of "village republics," producing the basic necessities of life—a type of cooperative structure where each family would purchase the necessities that it does not produce from other families, a relationship based on Gandhi's law of Swadeshi.[10] Gandhi was vehemently against the "soul-less machine" as well as the "soul-less state"; both enslave the individual and are not consistent with his Swaraj, self-government that is the basis of a just, nonviolent social order.[11] The touchstone or criterion for state action was provided by Gandhi in his advice to the ministers of a state government:

> Whenever you are in doubt, or when the self becomes too much with you, apply the following test. Recall the fact of the poorest and the weakest man whom you may have seen, and ask yourself if the step you contemplate is going to be of any use to *him*. Will he gain anything by it? Will it restore him . . . control over his own life and destiny? In other words, will it lead to Swaraj for the hungry and spiritually starving millions? Then you will find your doubts and . . . self melting away.[12]

The major instrument for the purpose of evolving such a social order was basic education and adult education. The main purpose of such education was to awaken the individual to "the sense of his dignity and power," make him understand his immediate environment and develop and sharpen his sense of perception and intelligence to *identify* the collective problems and seek their solution through "scientific inquiry" and intelligent effort.[13] An individual is the product of society and its past; he should, therefore, understand tradition, which is not possible without a certain reverence for the past—his tradition. But he need not be—should not be—a prisoner of the past; he has to develop a critical scientific temper to identify the problems tradition has created and revolutionize this tradition by adapting it creatively to the changing environment. Education has to develop such creative intelligence to meet life and its problems. "I am all for thoroughgoing, radical social reordering; but it must be an organic growth, not a violent superimposition," Gandhi said.[14]

His scheme of basic education emphasized learning through a craft. This was the only way to understand environment; further, creative intelligence develops through work by hand and through tackling problems of immediate relevance. Such a system would create a sense of dignity of labor and at the same time would be self-supporting, the financial means to be provided through the sale of the fruits of labor. In the Indian context, to be viable, such schools had to be self-supporting.[15]

Gandhi believed that such a school system would enable the students to understand their environment—problems of social order and techniques in agriculture, cottage industry, health, sanitation, and hygiene, housing, and village roads. The crux of technical change for Gandhi was in understanding the relevant problem. Once the problem was identified, the students would experiment to find a solution that was feasible. Once the problem was understood, and if its solution required outside assistance from research laboratories or scientific institutions (anywhere in the world), they would know whom to approach for assistance and how to ask and judge the relevance of the suggested solution. At all stages, the control of their environment would be—should be—in their hands. Such a system of education and research would be, according to Gandhi, "the spearhead of a silent social revolution fraught with the most far-reaching consequences."[16]

For Gandhi, the problem was to be understood and solved by immediate action. He was not in favor of imprisoning the future in the shackles of a long-term fixed plan; for him, attention was to be focussed on the immediate and the urgent. Once that was done with intelligence, the rest would follow. Hence his refrain: "One step enough for me"[17] and "sufficient unto the day is the evil thereof." Hence his emphasis on creative intelligence: "Swaraj is for the awakened, not for the sleepy and the ignorant." His nonviolent movement "is a revolution of thought, of spirit," of awakening each individual to the truth that he or she is "the guardian of his or her self-respect and liberty."[18]

Even during his lifetime, Gandhi strived ceaselessly to experiment with changes in technology suited to the Indian conditions. He improved the spinning-wheel and suggested to those qualified that they make further changes in it and enable it to operate with power. He set up all-India institutes to revive and foster cottage industry. He experimented with the use of human waste as organic manure for agriculture. He made original suggestions for the improvement of sanitation and hygiene and the construction of village roads. In health, he suggested and experimented with nature cures. His theory of education was translated into the concept of basic education. In all his endeavors for technological change, his criterion was simple: to devise such techniques and tools as villagers could make and afford and operate.[19]

For large-scale industry, education and research were to be related to immediate problems, and means to fund them were to be provided by the industry itself.[20] Once the whole effort in education and scientific-technical research was directed toward identifying and solving practical problems of immediate relevance, the nation would be able to afford "more and better libraries, more and better laboratories, more and better research institutes. . . . There will be truly original work instead of mere imitation. And the cost would be evenly and justly distributed."[21] The problem in India was immediate and unique: "The solution therefore must be original."[22]

THE SOCIALIST PERSPECTIVE

As with Gandhi, Nehru's vision of development policy was formed before India's independence.[23] He was influenced by the Marxian approach to history, Fabian socialism, and the Russian experiment in planned development. His conception of socialism, however, had the basic elements of Gandhi's moral approach. Gandhi influenced him in several ways. He approved of Gandhi's insistence on purity of means, a moral yet critical scientific approach to human problems, a concern for the poorest and the development of a socioeconomic order that had its roots in the living elements of Indian tradition and at the same time promoted world peace and cooperation.[24] Thus, a socialist pattern of society was to be evolved through democratic and peaceful means on the basis of and in tune with the national genius and cultural heritage.[25]

His major difference with Gandhi was with regard to development strategy. Unlike Gandhi, his approach to the problem of poverty and unemployment was indirect. His immediate concern was to initiate a process of planned, large-scale industrialization with emphasis on essential heavy industry based on modern science and technology; this would provide, according to him, a permanent and enduring solution to the poverty and unemployment problem.

India's poverty and backwardness in Nehru's mind were due to her failure to adopt modern technology and to adapt her socioeconomic structure to the modern scientific age. Hence a socialist pattern of society had to be evolved with modern technology. But this was possible only with the growing capacity (1) to produce machines that make machines, with the adoption of the latest—and hence more efficient—techniques and (2) to make original contributions to scientific and technical research. As Nehru stated:

> The three fundamental requirements of India, if she is to develop industrially and otherwise, are a heavy engineering and machine-making industry, scientific research institutes, and electric power. These must be the foundations of all planning. . . .[26]

> Political change there must be, but economic change is equally necessary. That change will have to be in the direction of a democratically planned collectivism. . . . A democratic collectivism need not mean an abolition of private property, but it will mean the public ownership of the basic and major industries. It will mean the cooperative or collective control of land. In India especially it will be necessary to have, in addition to the big industries, cooperatively controlled small and village industries.[27]

Unlike Gandhi, Nehru did not regard full employment as his immediate objective. It was necessary to strive for maximum employment, and for this purpose cottage and small industry had to be promoted, but only as a temporary means to relieve the unemployment problem. The traditional industries and techniques were obsolete and out of date and had to be supplanted by the "big machine." It was the modern sector that had to be developed; the traditional sector had to give way soon. Thus Nehru asserted:

> An attempt to build up a country's economy largely on the basis of cottage and small-scale industries is doomed to failure. It will not solve the

basic problems of the country or maintain freedom, nor will it fit in with the world framework, except as a colonial appendage.

Is it possible to have two entirely different kinds of economy in a country—one based on the big machine and industrialization, and the other mainly in cottage industries? This is hardly conceivable, for one must overcome the other, and there can be little doubt that the big machine will triumph unless it is forcibly prevented from doing so. Thus it is not a mere question of adjustment of the two forms of production and economy. One must be dominating and paramount, *with the other as complementary to it, fitting in where it can.* [Emphasis added.] The economy based on the latest technical achievements of the day must necessarily be the dominating one. If technology demands the big machine, as it does today in a large measure, then the big machine with all its implications and consequences must be accepted.... But, in any event, the latest technique has to be followed, and to adhere to out-worn and out-of-date methods of production, except as a temporary and stop-gap measure, is to arrest growth and development.[28]

In essence, technology for Nehru was a parameter, and its autonomous development had to dictate the socioeconomic structure. There s no appreciation of the role of technological research in *upgrading* traditional techniques, nor is there an awareness of the problem of *adapting* modern technology to India's circumstances and environment. In essence, the problem according to Nehru was that of cultivating adequate scientific and technological competence in order to *master* the latest techniques and conduct original research to advance the frontiers of scientific and technological knowledge; hence his emphasis on capital goods industry and scientific research.[29]

Thus, although Nehru starts, like Gandhi, with the human problems of poverty and unemployment, his emphasis subtly shifts from these basic problems to the problem of scientific and technological progress for its own sake. Gandhi's basic teaching appealed to Nehru; as he says: "In two respects the background of his thought had a vague but considerable influence; the fundamental test of everything was how far it benefited the masses, and the means were always important and could not be ignored even though the end in view was right, for the means governed the end and varied it."[30]

Until his death in 1964, Nehru was the Chairman of the National Planning Committee, constituted by the Congress in 1938. This Committee's findings had the stamp of Nehru's vision. As he said, "the National Planning Committee laid the greatest emphasis" on his "three fundamental requirements": heavy engineering and machine-making industry, scientific research institutes, and electric power.[31]

THE PERSPECTIVE OF FIVE-YEAR PLANS

After independence the interrupted work of the National Planning Committee was taken up by the Planning Commission, constituted in March, 1950. The National Planning Committee had started its work with the objective of providing an irreducible minimum standard of living for each individual. This was based on: (1) a balanced diet having a calorific value of 2400 to 2800 units for an adult worker, (2) 30 yards of clothing per year, and (3) a housing standard of 100 square feet. The achievement of these standards was estimated to cost about Rs. (rupees) 15–25 per capita per month (at 1938 prices) or at present prices about $15–25 per month—a standard to be attained in a decade.[32] It is worth noting that this approach, even as an exercise, was not taken up until the Third Plan (1961–65) when the national minimum income per month per individual was fixed at Rs. 20 per month at 1960–61 prices (or $8 at present prices)—a standard to be attained in fifteen years.[33]

By and large, the Planning Commission—again with Nehru as its chairman—adopted Nehru's strategy: "The central objective of planning in India at the present stage is to initiate a process of development which will raise living standards and open out to the people new opportunities for a richer and more varied life."[34] As with Nehru, so with the Commission, initiation of a process of development was the principal objective; with Gandhi, however, what mattered was the immediate objective of removing poverty through full employment. Planned development had to concentrate on socioeconomic and technological change. To quote the First Plan document: "The pace of economic development depends on a variety of factors which constitute the psychological and sociological setting within which the economy operates. . . . Basically, development involves securing higher productivity all round and this is a function of the degree of technological advance the community is able to make. The problem is not one merely of adopting and applying the process and techniques developed elsewhere, but of developing new techniques suited to local conditions. Modern technology is changing rapidly and no country can hope to maintain a steady pace of advance unless it keeps abreast of current developments."[35]

Thus the emphasis was on mastering the latest technology and new research, for which a variety of research institutes were set up during the First Plan period (1951–56).[36] For research results to ripen into actual technical innovations, it was recognized that pilot plant experiments might not suffice and that it might be necessary "to install semi-commercial or prototype plants to demonstrate effectively the new processes and their economic possibilities. . . . For this purpose, the establishment of a National Research Development Corporation of India has been proposed. . . ." Most of the research activity was to be carried out under the auspices of the Council of Scientific Research (CSIR), originally formed in 1942.[37]

There was no emphasis on *full* employment as a condition for removing poverty. "Maximum use of idle labor" was mentioned and it was indicated that "For absorbing all, or a large proportion, of the increase in working population each year in non-agricultural occupations, reliance will have to be placed mainly on small-scale and cottage industries involving comparatively small capital investment. If these industries are to afford employment at reasonable rates of wages, their technical efficiency will have to be increased and every effort will have to be made to see that the producer's reward is not intercepted by middlemen. . . . In fact, development is, in a sense, but another name for employment opportunities."[38]

The First Plan gave some recognition to the problem of upgrading technology in traditional industries: "The research institutions will bring the fruits of research within their reach and enable them to reduce costs and improve the quality of their products. The establishment of these laboratories and institutes is thus complementary to the objective of promoting the development of small-scale and cottage industries."[39] However, the emphasis was on research and not on the search for appropriate technology suited to solve the actual concrete problems of small industry. There was no technology policy for upgrading techniques in traditional industry, just as there was none for adapting modern technology. There was, in fact, no technology policy.

No plans subsequent to the first even discussed the technology question in the basic chapters outlining plan objectives and policies; "scientific research" was relegated to a separate chapter, more or less unrelated to the substantive parts of plan policies. And this chapter merely described the structure of the scientific research system and the research problems and results, without in any significant way relating them to the actual production or project problems. Scientific research was treated as an autonomous activity unrelated to the problems of development.

The significant part of the Second Plan (1957–61) concerned the strategy of industrial development, based on basic and heavy industry. Its two major objectives were: (1) "to attain a rapid growth of the national economy by increasing the scope and importance of the public sector and in this way to advance to a socialist pattern of society; and (2) to develop basic heavy industries for the manufacture of producer goods to strengthen the foundations of economic independence."[40]

Of course, there was emphasis on traditional skills and industries, which were expected to produce consumer goods to the required extent and at the same time serve as a stopgap device to provide maximum employment to the poorer section of the people. But there is no discussion of the problem of upgrading their technology; after the First Plan, this question is not even posed in any subsequent plan document.

Such, then, was the pattern of development policy in India. Scientific research had a place, but the plans lacked a technology policy. The scientific research system and the production system advanced on parallel lines, and there was no recognition of the need for a technology policy relating one to the other.

And so we come to the *abortive* Draft Fifth Plan document (1974—79), which in two crucial respects was a departure from the earlier plans. First, central importance was placed on the "removal of poverty and attainment of self-reliance" as "the major objectives that the country has set out to accomplish in the Fifth Plan."[41] Second, recognition was given to the need for a technology policy: "The Science and Technology plan, as an integral part of the Fifth Plan, is one of the major policy instruments for achieving the objective of self-reliance."[42] Not since the First Plan has the technology question been posed in the chapter on plan policies; here it appears under the title "Science and Technology Policy," a title never before used in the plan documents.

The Fifth Plan explicitly recognizes the lack of technology policy in the earlier plans. To quote:

> More importantly, perhaps, the structuring of the effort has been such that the goals of R&D programmes and projects have, often, not been derived directly from the technological needs of development projects, whether in industry or other areas. At the same time, the scope of "science planning" has not covered the whole of the "innovation chain." As a result, successful research results at the laboratory level have often failed to be linked to such elements as pilot plant work, design engineering, plant erection and commissioning and marketing, which are essential if the nation is to secure real and substantial, social and economic benefits from science and technology. . . . The task involved is really "planning the promotion of Science and Technology and their application to the Development and Security of the Nation. . . ." What we are primarily interested in is not a plan merely for education and research in various scientific and technological disciplines but a plan to harness science and technology for achieving the goals and programmes of the Fifth Plan.[43]

The Fifth Plan, thus, diagnoses the problem of technology policy, but it does not discuss with sufficient clarity and precision the crucial elements of a sound technology policy. It does not recognize, for instance, that these problems—ascertaining *relevant* research, *adapting* modern technology, and *upgrading* traditional technology—are interrelated and that their solutions crucially depend on the existence of adequate machinery to link the production and the technology systems.

BROAD OUTLINE OF A TECHNOLOGY POLICY

That India did not have and still does not have a sound technology policy as an integral and vital part of development strategy is not very surprising. Neither the development economists nor the international development agencies (like the World Bank) have appreciated the crucial importance of technology policy—and this in spite of Marx, Schumpeter, and Kuznets. The development economists have concentrated on methods and techniques of analysis rather than on the intelligent perception of the process of development and the human problem of poverty. In development policy, as in all walks of life, what matters is the creative perception of problems; logic and intellect have their role only if they are guided by creative intelligence and perception. Economists can afford to remain imprisoned in certain static molds and patterns of thought. But such unintelligence is fatal in policymakers and hence in development economists, for they have to deal with intelligent action in the present—in a world full of uncertainty and complexity—and not with thought, which is invariably related to the past.[44]

The less-developed countries with all their handicaps of a late start have one critical advantage: the accumulated and growing body of scientific and technological knowledge. But to use this knowledge creatively for solving their own problems, they need to develop competence in the field of technology choice through an institutionalized technology policy. Sophisticated techniques of planning or project evaluation are not—cannot be—a substitute for such a technology policy; the former in fact has relevance *only* with such a policy. What then are the major elements of a technology policy?

Obviously one has to start with the human problem of poverty, as Gandhi did, and try to solve it through means or instruments that are *directly* related to the problem. As Gandhi realized, it is neither practical nor desirable to provide a decent minimum of physical security to each family without providing it with work. So poverty can be removed only through full employment, the employment of each individual family. Starting from this focal point, and given the natural resources, the problem becomes very specific and concrete: how to choose organizational forms and technology so as to employ fully the available and growing labor force and at the same time ensure each family a decent minimum level of living initially and later a progressively rising minimum.[45]

Obviously wherever one wants to go one has no alternative but to start from where one is. In this context one has to start from the traditional sector with its skills, techniques, and organization. In India (and probably in many other less-developed countries), this sector employs about 80 percent of the labor force (60 percent in agriculture and related activities and 20 percent in traditional industry like hand-loom weaving, gur-making, rural transport, agriculture, dairy and cattle breeding, forestry, poultry farming, piggeries, fisheries, organic manure, agricultural implements, and household utensils and appliances); about 10 percent is employed in the modern sector and the remaining 10 percent is actually underemployed or unemployed. With growth of population at the rate of 2.5 percent per annum, the modern sector, even if it grew at 20 percent in terms of employment, would not be able to provide employment to the full extent even for the increased labor force—let alone for the presently unemployed. Thus it is obvious that the traditional sector would have to absorb the currently unemployed as well as a part of the increase in labor force.

Hence, as Gandhi realized, the crucial problem is that of promoting the viable growth of the traditional sector and organically linking the modern sector growth with that of the traditional in a manner that supplements and reinforces the growth and development of the latter. It is not possible to supplant the traditional skills and techniques by the modern ones; that cannot solve the poverty problem.

In terms of technology choice, then, there are two problems: the problem of upgrading the traditional technology and the problem of adapting and improving modern technology in a manner that is consistent with development objectives, changing institutional structure, and natural resources. Without such upgrading and adapting, it is impossible to provide a rising minimum level of living to each family. For both purposes, it would be essential to identify the concrete ideas relating to the projects or project complexes that have to be undertaken. Technology choice problems can arise only after such initial identification. For both the sectors—traditional and modern—this identification process needs to be institutionalized in, say, Project Identification Centres (PICs).

With such identified project ideas should begin the search for appropriate technology. And this function is of critical significance for both the problem of upgrading and the problem of adapting and improving. This is the design and engineering function that needs to be institutionalized in Technical Consultancy Service Centres (TCSCs).

It is the function of such a TCSC to design the machines, processes, and products. It requires knowledge of the specific initial conditions of both the production and the organizational structures. It requires identification of current knowledge and the current ways in which it can be applied. Once one starts experimenting with the design of projects or project complexes that are to be implemented—and this is how the design-engineering function is crucially linked with that of the PICs and the financial system (FS)—one has set in motion a process of search for relevant knowledge as well as relevant research, a cumulative process of innovation that is self-reinforcing.

Starting from a project idea, a relevant TCSC could (1) help in making appropriate technological choices and in diffusing relevant technology, (2) function as a vehicle for absorbing relevant modern technology by serving as an effective communication link with foreign sources of technology, (3) support and improve machine-building capacities by providing machine industries with designs as well as links with the production structure, (4) identify technological research problems and thus link research with industry, (5) generate a wide variety of project ideas in diverse fields, and (6) help the PICs by providing them with norms for input, skill and capital coefficients. Thus a TCSC can serve as a focal point for generating a self-reinforcing cumulative process of innovation, consistent with a country's development strategy.

Such TCSCs, however, have to be functionally linked with PICs and the FS; without such an organic functional link, they would become obsolete and operationally irrelevant. Again, they have to be organically linked with the Technology Research Centres (TRCs); without this link technological research would become irrelevant. A research organization is essential; but it can derive its meaning and purpose only if it is related to the design-engineering function that is intimately and vitally related to actual projects or project complexes, which again have to be an integral part of a development strategy.

Strategy, project ideas, design and engineering (search for appropriate techniques), financial systems, project implementation, new project ideas: this is one sequence that organically relates one function to the other. The second organically related sequence is: strategy, project ideas, design and engineering, technological research, design engineering, financial system, project implementation, new project ideas. The third such sequence is: design engineering; product, process, and machine ideas; project ideas; design and engineering; financial system; project implementation; new project ideas. The important point to realize, however, is the central and crucial significance in these chains of the design and engineering function. In the field of agriculture (e.g., high-yielding varieties of seeds), there is some appreciation of the significance of these organically related functions; but in other areas this obvious and more or less self-evident function of technology policy is still not understood.

What is the role of the international community in general and of international development institutions (like the World Bank) in particular in this field? First they can urge the developing countries to set up PICs, TCSCs, and TRCs in relevant fields, link them with the financial system and assist them in the initial stages in a variety of ways; second, for certain complex fields, they can set up regional/international TCSCs; third, they can start on their own international/regional TRCs which should work primarily on research problems identified by the national/regional TCSCs, but which should also function as clearing-houses for relevant information for the national TCSCs and TRCs. It is only thus that national and international efforts can be effectively integrated for solving the urgent and immediate problem of poverty and socioeconomic development.

SOME CASE STUDIES IN TECHNOLOGY CHOICE

Even now there is no clarity about the role of technology policy in the development process in India. It is, of course, realized that India lacked such a policy in the past, but the Planning Commission has not been able to formulate a concrete institutionalized technology policy as an integral part of its development strategy.

However, even in the past, there were some concrete cases in which the organic functional sequences discussed in the previous section did operate with the expected results. The Swaraj tractor case shows with clarity the soundness of this sequence; it also shows how lack of clarity with regard to the technology choice problem led the central government policy-makers (including the apex research institution) to distrust indigenous technology efforts. The Bokaro steel case illustrates the implications of the neglect of the crucial function of design and engineering in technology policy as well as the power struggle among the elite and its inferiority complex vis-à-vis foreign technology.

The other two cases—the Bio-gas Project and the Bamboo Tube Well—illustrate the Gandhian institutions that have been kept alive and supported by the government partly out of respect for him and partly because of the compulsions of the actual situation. In spite of the large *potential* impact on poverty and employment of these two projects, it is surprising to find that the Fifth Plan hardly mentions them, let alone deriving any lessons with regard to technology policy for the traditional sector. The second shows the significance of Gandhi's strategy of basic education and at the same time the typical attitude of the educated urban elite in the government.

Swaraj Tractor Project[46]

For the small farms in India it was essential to have low-horsepower, multi-purpose tractors suited to their needs as well as resources. The multi-national companies were producing tractors of 30 h.p. and above, using obsolete technology. Russian assistance was sought in 1965, but this implied a large foreign exchange cost and a large number of Russian experts. At any rate, the Russians suggested the purchase of 20 h.p. tractors from Czechoslovakia. Mr. Suri, the Director-in-Charge of the Central Mechanical Engineering Research Institute (CMERI), convinced the planners that India had the capacity to design, engineer, and manufacture a 20 h.p. tractor suited to the specific Indian conditions. The Swaraj tractor was thus devised, a tractor which has passed the tests of the Tractor Testing and Training Station (TTTS). Only two other tractors—both imported—have passed this test.

The Ministry of Agriculture and the public-sector Hindustan Machine Tools (HMT) still favored Zeteor, a 20 h.p. tractor of Czech design, on the grounds that Zeteor was a production model, while Swaraj was still a prototype. HMT decided to produce Zeteor on a turnkey contract with Czechoslovakia. The Swaraj tractor, thus, was left without a promoter. Neither the Central Government nor the apex scientific research body, the Council of Scientific and Industrial Research (CSIR), supported this project; yet they did support with foreign collaboration several other projects that were much less efficient than the Swaraj tractor.

Fortunately for the Swaraj tractor, the Punjab Industrial Development Corporation was attracted to the idea. The design-engineering tasks were offered to Mr. Suri, who was responsible for the tractor design, and to his consulting firm. The technical personnel were drawn from CMERI staff who had worked on the design. The financing problem was solved at the initiative of the Industrial Development Bank of India (IDBI), which appraised the project, was convinced of its viability and was attracted by the fact that the tractor was indigenously designed and that its manufacture was based on skills, materials and equipment available in India. The IDBI supported the project also on the grounds of the project's vital and organic links with small-scale ancillary industries, from which the Swaraj tractor project was to buy a substantial number of parts. This project has been successfully implemented.

Bokaro Steel Plant[47]

For reasons of resource endowment and demand pattern, steel production in India was an obvious choice. The Bokaro steel plant was to be the fourth integrated steel plant in the public sector and the sixth in India as a whole. The idea originated in 1955, only a little later than those that led to the first three public-sector steel plants built on a turnkey basis with German, Russian and British assistance respectively.

The Bokaro steel plant, however, was to be built on the basis of Indian design, engineering, and construction skills. For the purpose, a deliberate attempt was made to develop Indian design-engineering skills. Dr. Dastur, an Indian by birth and an American citizen, was persuaded by the then Finance Minister, T.T. Krishnamachari, to form a consulting firm in India in 1954. Thus was set up the steel consultancy firm, M.N. Dastur and Company (Dasturco), which was soon to establish an international reputation. This firm was assigned the consulting role for Bokaro at the insistence of the Prime Minister, Jawarharlal Nehru, in 1957.

The decision to build Bokaro on the basis of Indian design and engineering skills was taken very rationally in the light of past experience with turnkey projects. There were definite advantages in having structural items designed by Indian consultants, who were aware of Indian fabricating needs and limitations. The heavy design features of steel plant equipment, though technically sound, are not necessarily appropriate or economical for Indian conditions, as they have been based in most cases on the codes and practices evolved for specific conditions prevailing in foreign countries. The heavy equipment design, in turn, necessitates heavier structures, foundations, materials-handling facilities, etc., which again add to the steel plant costs. There was thus a specific need for an Indian initiative in the development of design norms for its own steel plants.[48]

By 1959 Dasturco finalized their preliminary report, which was thoroughly scrutinized by the Technical Committee of Hindustan Steel Limited and commended by the Russian and American steel experts who had seen it. In 1962, Dasturco was commissioned to prepare the detailed project report, while at the same time the U.S. Steel Corporation was asked by the U.S. Agency for International Development to make a feasibility study. Both reports were submitted in 1963.

It is interesting to compare the cost and other estimates of the two reports (see Table 1). In spite of the greater input of capital, foreign skill and foreign exchange in the U.S. Steel Corporation version of the project, its technology was to be inferior; it was of the semi-continuous casting variety, instead of the more advanced and efficient method of continuous casting of flat steel items, recommended by Dasturco. Again the provision of consulting, management and designing fees by the U.S. Steel Corporation was $190.4 million or as much as 15 percent of the total plant cost estimate.

Since the U.S. Steel negotiations broke off, Dasturco was appointed as consulting engineer for the Bokaro project starting in April 1964. Meanwhile, the Russians showed willingness to finance Bokaro on a turnkey basis and submitted their project report in December 1965. Dasturco was kept out of the basic part of the project but was asked to comment on the Russian report—within seven weeks.

Dasturco found that the cost of the project, as estimated by the Russians, could be reduced by Rs. 1,075 million and that the technology could be considerably improved. Dasturco argued for installing large-sized converters for the steel-melting shop on grounds of compelling advantages of economies of scale; this in itself would have meant a cost-saving of Rs. 187.2 million. Furthermore, the Soviet technology was found to be obsolete; it did not adopt the continuous casting method, which eliminates the need for a slabbing mill. Again Dasturco found the provision for the services of foreign consultants and specialists (450 to 500 Soviet experts) to be excessive. They argued that the bulk of the work for which such large amounts were to be paid (including foreign exchange) was essentially local work and could be competently undertaken at a relatively low cost by the Indian side, which has greater familiarity with Indian condi-

Table 1. Comparative Cost Estimates.

	Dasturco	*U.S. Steel Corporation*
Capacity	1.5 million tons	1.4 million tons
Project cost	$751.47 million	$919.428 million
Foreign exchange cost	$318.48 million	$512.588 million
Date of completion	1969	1971
Foreign experts	30–40	670

tions and practices. The Russians accepted a reduction of only Rs. 95 million, as against the total cost reduction suggested by Dasturco of Rs. 1,075 million, from the total estimated project cost of Rs. 7,700 million for 4 million tons capacity. Again the Russians stuck to the obsolete technology, even though they recognized the superiority of the technology suggested by Dasturco. The Government of India, however, accepted the Russian turnkey project. This is all the more amazing in light of the fact that the Indians started with a clear vision and had the requisite design and engineering skills—skills which could have reduced the cost of the project, improved upon it, and thus set the stage for initiating a process of technology self-reliance in a basic field like steel.

The Indian decision cannot be understood simply in terms of Russian aid. The foreign exchange cost would not have been more than $64 million per year, even with an assumed gestation lag of seven years. This alone would not have been a sufficiently compelling reason for not utilizing Dasturco, for whose extensive services a six-year contract had already been signed. A more plausible reason would seem to have been anxiety on the part of the Ministry of Steel lest Dasturco's increasing influence should detract from the Ministry's own status in the power hierarchy. This power struggle was clothed in arguments of an ideological character—developing public-sector consultancy skills and the like—and the ostensible argument used was the foreign exchange assistance to be obtained from the Russians on Russian terms. Why then did the Government of India not function in the country's interests, and why did it succumb to the penny-wise-pound-foolish logic of the Ministry of Steel? Again, why did it not bargain hard with the Russians and emphasize the overriding need to build a steel plant based primarily on Indian design, engineering, and construction skills—the concept upon which the Bokaro project idea had been formulated by Prime Minister Nehru and the purpose for which Dr. Dastur had been encouraged to start a consultancy service in India? Again, the answer, under the circumstances, can only be a lack of clarity with regard to development strategy and objectives and a sense of inferiority, making for an imitative psychology among the members of the power elite.[49]

Bio-gas Project [50]

The concept behind this project is to use waste material as fuel in the production of both energy and fertilizer on viable small-scale farms. (Since cow-dung—gobar—is used as the fermentable material, these projects are called in India gobar-gas projects.) A classic paper on the fermentation of cellulosic waste was published in 1923 by the

Indian Institute of Science, Bangalore. The purpose of this paper was to contribute to a solution of the energy problem in India. It was followed by design-engineering work, and designs of various types of small bio-gas plants were accomplished in India. Indian work on the subject was in the forefront until 1952, after which the imitative elite neglected this work in favor of rural electrification along conventional lines and fertilizer production in large naphtha-based plants.

This development is a sad commentary on the Indian power elite and its norms and standards, for, as Table 2 indicates, gobar-gas plants can produce both fertilizers and energy cheaper than the conventional modern technology-based projects.

In spite of its obvious advantages, this indigenous technology was neglected until 1970. However, interest in this work was kept alive by authentic, but powerless, elite groups like the Khadi and Village Industries Commission, the Indian Agricultural Research Institute, and the Gobar-Gas Research Station. As a result, twenty different sizes and capacities of gobar-gas plants have now been designed and 8,000 such plants are in operation. With the oil crisis, this technology has been recognized and it is planned to build 50,000 such plants in the next few years—an effort that is still much smaller than that applied to modern fertilizer plants. At any rate, the design-engineering skills already available have identified several areas for research that can potentially reduce costs still further and link gobar-gas plants organically to agricultural development based on small farms.

The Bamboo Tube Well [51]

This is a fascinating story of indigenous technology development by a reasonably educated middle-sized farmer in the Bihar state. As with all farms in this area, irrigation was a problem. And for the new high-yielding varieties of seeds, irrigation was essential. Mr. Ram Prasad decided to employ tube-well irrigation and borrowed money for the purpose in July 1968. His failure to obtain iron pipe gave him the idea of using a bamboo tube well. He succeeded in this experiment and within less than a month, his bamboo tube well began to function. Within the next six months, some farmers took to this technology and by March 1973, more than 33,000 bamboo tube wells were sunk in ten districts of Bihar. The cost of such a tube well is only 10 percent of the cost of sinking an iron tube well to the same depth.

Of course, Ram Prasad received a letter of congratulations from the State Ministry for Agriculture in recognition of the worth of his invention. Nonetheless, the technology policy of the government has

Table 2. Implications of Production of 230,000 Tons Nitrogen.

	Large-scale Coal-based Fertilizer Plant	Bio-gas Fertilizer Plant
Number of units	1	26,160
Capital cost	1,200 Rs.	1,070 Rs.
Foreign exchange cost	500 Rs.	Nil
Employment	1,000	130,750
Energy	About 35 MW consumption	6,350,000 MWH generation

taken no note of this indigenous technology; the science and technology elite has not given attention to this as it has not to several such projects. It is surprising that even the economists have not studied these efforts.

Traditional industries like handloom weaving, gur-making, and rural transport provide the largest employment next to agriculture. It is surprising to find that these industries and their technical structure are not even carefully analyzed; there is little concern with upgrading their organizational and technical base. The educated elite not only has an urban bias; it has a Western bias, and it is mistaking Westernization for modernization. Modernization, as Gandhi and Nehru understood it, implies the use of creative intelligence—not merely intellect—to identify and solve problems as they emerge. This implies *understanding* the tradition to change it radically from within.

NOTES TO CHAPTER EIGHT

1. See The New Testament: "Take *that* thine *is*, and go thy way: I will give unto this last, even as unto thee." (Matthew 20:14). Sarvodaya conveys that the *good of all is served by promoting the good of the poorest—the lowliest and the lost.* See Nirmal Kumar Bose, *Selections from Gandhi* (Ahmedabad: Navjivan Publishing House: April 1957), pp. 38–41.

2. See Charan Singh, *India's Economic Policy: The Gandhian Blueprint* (New Delhi: Vikas Publishing House, 1978).

3. See V.V. Bhatt, *Two Decades of Development: The Indian Experiment* (Bombay: Vora, 1973), ch. 4.

4. See Bose, *Selections from Gandhi.* To quote Gandhi: "True to his poetical instinct, the poet (reference is to Tagore) lives for the morrow and would have us do likewise. He presents to our admiring gaze the beautiful picture of the birds early in the morning singing hymns of praise as they soar into the sky.

These birds have had their day's food and soared with rested wings, in whose veins new blood had flown during the previous night. But I have had the pain of watching birds who for want of strength could not be coaxed even into a flutter of their wings. The human bird under the Indian sky gets up weaker than when he pretended to retire. For millions it is an eternal vigil or an eternal trance. It is an indescribably painful state which has got to be experienced to be realized. I have found it impossible to soothe suffering patients with a song from Kabir. The hungry millions ask for one poem—invigorating food. They cannot be given it. They must earn it. And they can earn only by the sweat of their brow." Again: "Imagine, therefore, what a calamity it must be to have 300 millions unemployed, several millions becoming degraded every day for want of employment, devoid of self-respect, devoid of faith in God. I may as well place before the dog over there the message of God as before those hungry millions who have no lustre in their eyes and whose only God is their bread. I can take before them a message of God only by taking the message of sacred work before them. . . . To them God can only appear as bread and butter." Further: "My *ahimsa* would not tolerate the idea of giving a free meal to a healthy person who has not worked for it in some honest way and, if I had the power, I would stop every *Sadavrata* where free meals are given. It has degraded the nation and it has encouraged laziness, idleness, hypocrisy and even crime . . . Do not say you will maintain the poor on charity (pp. 46-47).

5. Ibid., pp. 48-55.

6. See my essay "On Development Problem and Strategy," in *Society and Change*, ed. K.S. Krishnaswami et al. (Bombay: Oxford University Press, 1977).

7. George Woodcock, *Mohandas Gandhi* (New York: Viking, 1971), ch. 7. See also Bose, *Selections from Gandhi*. To quote Gandhi: "My idea of village Swaraj is that it is a complete republic, independent of its neighbours for its vital wants, and yet interdependent for many others in which dependence is a necessity. Thus every village's first concern will be to grow its own food crops and cotton for its cloth. It should have a reserve for its cattle, recreation and playground for adults and children. Then if there is more land available, it will grow *useful* money crops, thus excluding *ganja*, tobacco, opium and the like. It will have its own waterworks ensuring clean supply. This can be done through controlled wells and tanks. Education will be compulsory up to the final basic course. As far as possible every activity will be conducted on the cooperative basis. . . . non-violence with its technique of *Satyagraha* and noncooperation will be the sanction of the village community. There will be a compulsory service of village guards who will be selected by rotation from the register maintained by the village. The government of the village will be conducted by the *Panchayat* of five persons, annually elected by the adult villagers, male and female, possessing minimum prescribed qualifications. . . . Here there is perfect democracy based upon individual freedom. The individual is the architect of his own government. . . . He and his village are able to defy the might of the world. For the law governing every villager is that he will suffer death in the defence of his and his village's honour" (pp. 73-74). See also Woodcock, *Mohandas Gandhi*. To quote Gandhi: "In this structure composed of innumerable villages . . . life will not be a pyramid with the apex sustained by the bottom. But it will be an oceanic

circle whose centre will be the individual always ready to perish for the village, the latter ready to perish for the circle of villages, till at last the whole becomes one life composed of individuals . . . . The outermost circumference will not wield power to crush the inner circle but will give strength to all within and derive its own strength from it" (p. 94).

8. To quote Gandhi: "If I can convert the country to my point of view, the social order of the future will be based predominantly on the *Charkha* and all it implies. It will include everything that promotes the well-being of the villagers. I do visualise electricity, ship-building, ironworks, machine-making and the like existing side by side with village handicrafts. But the order of dependence will be reversed. Hitherto, the industrialisation has been so planned as to destroy the villages and the village crafts. *In the State of the future it will subserve the villages and their crafts.* I do not share the socialist belief that centralisation of the necessaries of life will conduce to the common welfare, that is, where the centralised industries are planned and owned by the State" (quoted in Singh, p. 53).

9. To quote Gandhi: "Indeed at the root of this doctrine of equal distribution must lie that of the trusteeship of the wealthy for superfluous wealth possessed by them. For, according to the doctrine, they may not possess a rupee more than their neighbours. How is this to be brought about? Non-violently? Or should the wealthy be dispossessed of their possessions? To do this we would naturally have to resort to violence. This violent action cannot benefit society. Society will be the poorer, for it will lose the gifts of a man who knows how to accumulate wealth. Therefore the non-violent way is evidently superior. The rich man will be left in possession of his wealth, of which he will use what he reasonably requires for his personal needs and will act as a trustee for the remainder to be used for the society. In this argument, honesty on the part of the trustee is assumed."

"If however, in spite of the utmost effort, the rich do not become guardians of the poor in the true sense of the term . . . what is to be done? In trying to find out the solution of this riddle I have lighted on non-violent non-cooperation and civil disobedience as the right and infallible means. The rich cannot accumulate wealth without the cooperation of the poor in society. If this knowledge were to penetrate to and spread amongst the poor, they would become strong and would learn how to free themselves by means of non-violence from the crushing inequalities which have brought them to the verge of starvation . . . all the capitalists will have an opportunity of becoming statutory trustees. But such a statute will not be imposed from above. It will have to come from below" (quoted in Bose, pp. 78–80).

10. "The broad definition of Swadeshi is the use of all home-made things to the exclusion of foreign things, in so far as such use is necessary for the protection of home-industry, more especially the industries without which India will become pauperised . . . But even Swadeshi, like any other good thing, can be ridden to death if it is made a fetish. That is a danger that must be guarded against. To reject foreign manufactures merely because they are foreign, and to go on wasting national time and money in the promotion in one's country of

manufactures for which it is not suited would be criminal folly, and a negation of the Swadeshi spirit . . . Swadeshism is not a cult of hatred. It is a doctrine of selfless service, that has its roots in the purest *ahimsa*, i.e., love" (quoted in Bose, pp. 306–307).

11. "I look upon an increase in the power of the State with the greatest fear because, although while apparently doing good by minimising exploitation, it does the greatest harm to mankind by destroying individuality which lies at the root of all progress . . . The State represents violence in a concentrated and organised form. The individual has a soul, but as the State is a soulless machine, it can never be weaned from violence to which it owes its very existence. . . .

"It is my firm conviction that if the State suppresses capitalism by violence, it will be caught in the coils of violence itself and fail to develop non-violence at any time . . . What I would personally prefer, would be, not a centralisation of power in the hands of the State but an extension of the sense of trusteeship; as in my opinion, the violence of private ownership is less injurious than the violence of the State. However, if it is unavoidable, I would support a minimum of State-ownership" (see Bose, pp. 40–41).

12. Quoted in Bhatt, *Two Decades of Development*, p. 1.

13. Bose, *Selections from Gandhi*, pp. 283–298.

14. Ibid., p. 131.

15. See Gunnar Myrdal, *Asian Drama*, vol. 3 (New York: The Twentieth Century Fund, 1968), pp. 1655–1658. See also my *Education and Economic Development* (Bombay: Vora, 1972), pp. 8–10.

16. See Bose, *Selections from Gandhi*, pp. 291–297.

17. See Woodcock, *Mohandas Gandhi*. To quote Woodcock: "The 'one-step-enough-for-me' approach was in complete accordance with the general philosophy of *Satyagraha* and also with the teachings of the Bhagavat Gita as Gandhi interpreted them. . . . its exponents developed these principles experimentally in the world of action, where everything is relative and mutable, and in this sense the Gandhian philosophy is one of endless becoming, which makes it— in a somewhat different way from left-wing Marxism—a doctrine of permanent revolution. No society can ever be perfect . . . and for this reason the *Satyagrahi* cannot accept the rigid plans of the Utopian theorist, since he has to be sensitive to the constantly changing demands of human relations, and these assail him, not in the future, but now and here" (pp. 86–87).

18. Bose, p. 32, 43.

19. Ibid., pp. 62–89; see also Woodcock, ch. 7.

20. Bose, pp. 292–298.

21. Ibid., p. 95, 297.

22. Ibid., p. 55.

23. Gandhi's basic ideas on Sarvodaya were formulated as early as 1909, Nehru's as early as 1926. See Jawaharlal Nehru, *The Discovery of India* (Bombay: Asia Publishing House, 1969).

24. Ibid., pp. 356–65.

25. Nehru used to quote Gandhi approvingly on this. To quote Gandhi: "I do not want my house to be walled in on all sides and my windows to be

stuffed. I want the cultures of all lands to be blown about my house as freely as possible. But I refuse to be blown off my feet by any . . . . Mine is not a religion of the prison-house." See Bose, p. 298.

26. Nehru, *The Discovery of India*, p. 410.

27. Ibid., p. 522.

28. Ibid., pp. 404-409.

29. Apart from scientific research, what Nehru emphasized, as a way of life, was the scientific temper. "It is the scientific approach, the adventurous and yet critical temper of science, the search for truth and new knowledge, the refusal to accept anything without testing and trial, the capacity to change previous conclusions in the face of new evidence, the reliance on observed fact and not on preconceived theory, the hard discipline of the mind—all this is necessary, not merely for the application of science but for life itself and the solution of its many problems. Too many scientists today, who swear by science, forget all about it outside their particular spheres. The scientific approach and temper are, or should be, a way of life, a process of thinking, a method of acting and associating with our fellowmen. . . . It is the temper of a free man" (see Nehru, p. 512).

This was precisely the approach of Gandhi; he called his life "experiments with truth" (see Bose, pp. vi-x). But Gandhi's emphasis was on intelligence, perception rather than *only* on intellect. See J. Krishnamurti, *The Awakening of Intelligence* (New York: Harper & Row, 1973), on this. To quote: "May we say, thought is barren . . . which is mechanical and all the rest of it? Thought is a pointer, but without intelligence the pointer has no value . . . so intelligence is necessary. Without it thought has no meaning at all . . . . As we begin to inquire into it, or in inquiring, we come to this source. Is it a perception, an insight, and has that insight nothing whatsoever to do with thought? Is insight the result of thought? The conclusion of an insight is thought, but insight itself is not thought . . . you come upon it when you see the whole thing. So insight is the perception of the whole . . . the quality of mind that sees the whole is not touched by thought; therefore, there is perception, there is insight" (Ibid., pp. 520-35).

Compare Gandhi: "Truly beautiful creation comes when right perception is at work. If these moments are rare in life, they are also rare in art" (Bose, p. 304). Again: "In dealing with living entities, the dry syllogistic method leads not only to bad logic but sometimes to fatal logic. For if you miss even a tiny factor—and you never have control over all the factors that enter into dealings with human beings, your conclusion is likely to be wrong. Therefore, you never reach the final truth, you only reach an approximation; and that too if you are extra careful in your dealings" (Bose, p. 44).

30. Nehru, p. 362.

31. Ibid., p. 410.

32. Ibid., pp. 397-98.

33. See T.N. Srinivasan and P.K. Bardhan, eds., *Poverty and Income Distribution in India* (Calcutta: Statistical Publishing Society, 1974), pp. 9-39.

34. Planning Commission, Government of India, *The First Five-Year Plan* (New Delhi: 1952), p. 7.

35. Ibid., pp. 12-13.

36. Ibid., ch. 28.

37. Ibid., p. 418.

38. Ibid., p. 25.

39. Ibid., p. 413.

40. Planning Commission, Government of India, *Papers Relating to the Formulation of The Second Five-Year Plan* (New Delhi: 1955), p. 42.

41. Planning Commission, Government of India, *Draft Fifth Five-Year Plan—1974-79* (New Delhi: 1974), p. 1.

42. Ibid., p. 18.

43. Ibid., pp. 216-18.

44. See my article, "On Technology Policy and Its Institutional Frame," *World Development* 3, no. 9 (September 1975): 651-63.

45. On this problem see my essay "On Development Problem and Strategy."

46. The facts about this case are taken from my report, *Decision-making in the Public Sector: A Case Study of Swaraj Tractor*, Domestic Finance Studies No. 48 (World Bank, Development Economics Department, February 1978).

47. For details about this project, see Padma Desai, *The Bokaro Steel Plant* (Amsterdam: North Holland, 1972).

48. See M.N. Dastur, "New Strategy for India's Steel Development," *Economic and Political Weekly* 3, nos. 31-33 (1972).

49. Nehru himself was quite conscious of the nature of this imitative psychology of the elite: "It is not only the Communist Party in India that has failed in this respect. There are others who talk glibly of modernism and modern spirit and the essence of Western culture, and are at the same time ignorant of their own culture. Unlike the Communists, they have no ideal that moves them and no driving force that carries them forward. They take the external forms and outer trappings of the West (and often some of the less desirable features), and imagine that they are in the vanguard of an advancing civilisation. Naive and shallow and yet full of their own conceits, they live, chiefly in a few large cities, an artificial life which has no living contacts with the culture of the East or of the West," Nehru, p. 517. See also, S.N. Ganguly, *Tradition, Modernity and Development* (New Delhi: Macmillan of India, 1977).

50. For details about the economies and engineering of gobar-gas plants, see C.R. Prasad, K. Krishna Prasad, and A.K.N. Reddy, "Bio-Gas Plants: Prospects, Problems and Tasks," *Economic and Political Weekly* 9, no. 32 (1974); and A.K.N. Reddy and K. Krishna Prasad, "Technological Alternatives and the Indian Energy Crisis," *Economic and Political Weekly* 12, nos. 33-34 (1977).

51. For details, see Arthur J. Dommen, "The Bamboo Tube Well: A Note on an Example of Indigenous Technology," *Economic Development and Cultural Change* 23, no. 3 (April 1975): 483-89.

✳ *Chapter Nine*

China's Experience with Rural Small-Scale Industry

Dwight H. Perkins

For twenty years the People's Republic of China has experimented with the development of small-scale industrial enterprises in both rural and urban areas. Small-scale industries, of course, were not new to China in the 1950s. Industrial products in traditional Chinese society were largely supplied by a handicraft sector scattered the length and breadth of the country, mainly in small units. Despite competition from imports of the advanced industrial nations, these traditional enterprises continued to dominate the food processing industry and much of the clothing industry even in the 1950s. What was different about the small-scale industrial program that began in the late 1950s was the attempt to discover alternative technologies for industrial sectors usually dominated by modern large-scale units.

China's initial move into a small-scale industrial program, like many of the other experiments of the Great Leap Forward of 1958–60, was not a success. Small-scale enterprises were established in a great many industries, but it was in iron and steel that the big push was made. Virtually every school, factory, and commune was expected to have its backyard furnace. Scrap iron and steel were to be found to feed these furnaces, which, it was hoped, would then supply everything from manufactures of simple farm implements to large modern machinery plants with their steel requirements. The result, as is now well known, was misuse of labor in all sectors, a frantic search for scrap that sometimes led to the melting down of useful implements in order to meet quotas, and the production of large quantities of iron and steel of such low quality that it was often no better than scrap.

By the mid–1960s, however, efforts began to resuscitate small-scale industry, but on a more cautious and carefully calculated basis. No longer were small-scale enterprises seen as an across-the-board answer to the problem of how to accelerate China's industrial development. Instead, the new effort centered on the rural areas and was part of a more general program to accelerate agricultural development. State planners realized that rural China could not do the job of raising agricultural production alone. Modern input, from chemical fertilizer to cement to electricity, would have to play a role.

This change in thinking was due in part to China's unfortunate experience during the Great Leap Forward. In the 1950s the Chinese leadership had hoped that the rural sector could pull itself up by its own bootstraps. First cooperatives and then communes were to mobilize great quantities of underemployed rural labor to build irrigation systems, roads, and other forms of capital construction. Inputs from the modern sector in the form of machinery or steel were to be held to a minimum. Wherever possible, work was to be accomplished with homemade tools and strong backs. The result: a massive outpouring of effort leading to a sharp decline in farm output. In the early 1960s the communes were reorganized with the production team, a small subunit of the commune, as the basic unit in the countryside. This effective organizational unit was to be the recipient of inputs from the modern sector.

When the initial decision to accelerate the supply of modern inputs to the rural areas was made, there still existed the problem of who was to supply those inputs. Imports from abroad were one possibility, but China was a large country and like most large countries its exports, and hence its ability to earn foreign exchange to pay for imports, were limited. Eventually a large percentage of agriculture's requirements would have to be met from domestic sources, but the questions remained, what should the scale of those domestic enterprises be and where should they be located? At first many of the new plants producing agricultural inputs were large in scale, and some were imported. But increasingly in the 1960s and early 1970s, the nation turned to small-scale units located in the countryside. By the mid–1970s Chinese planners were once again importing large plants from abroad, particularly in the chemical fertilizer sector. At the same time, rural small-scale plants, far from being abandoned, continued to grow.

China's rural small-scale industry program, therefore, arose out of the perceived requirements of the rural sector. As I shall attempt to demonstrate in this essay, small-scale plants were a rational way to meet these requirements, even when "rational" is defined in purely economic terms. If these enterprises had not been economical, they would probably have fared little better than the backyard iron and steel furnaces. However large the social benefits of such industries to rural society, it is doubtful that Chinese leaders would have continued to push them on such a scale if larger urban plants could have delivered the same inputs at a substantially lower cost.

Social values nevertheless did have a major impact on rural small-scale industry development, and that development in turn had an impact on rural society. In the most fundamental sense, what made rural industries economical was the decision of China's leaders to make rural development a centerpiece of their overall development program. If, instead, the leadership had continued to emphasize machines and steel to make more machines and steel as in the 1950s, it is doubtful if the rural industrial program would ever have gotten off the ground. Let us consider now how China's overall rural development strategy, when combined with underlying conditions in China's countryside, made rural small-scale industry economical. We shall then return to questions of the social impact of this program and the degree to which the Chinese experience is transferable.

SOURCES OF DEMAND

Rural industries of any scale are not going to appear unless a demand for their product already exists. Only rarely will such industries be built ahead of demand. In the Chinese case it was the massive rural construction effort that accounted for the initial rise in demand for cement, the first of the new type of rural industries to really take off. The cooperatives and communes could readily mobilize the labor required for most projects and with only the simplest of tools that labor could move large amounts of rock and dirt. But dirt-filled dams require outer walls of cement, and irrigation channels and reservoirs are more effective if lined with cement. The cement does not have to be of a particularly high quality to still be of use. The Chinese talk about their ability to make Mark 500 cement in small-scale plants, but generally their cement quality is well below that level. Such cement would be of little use in twelve-story reinforced concrete apartment buildings, but no such buildings are being constructed in the Chinese countryside.

Once rural construction has proceeded to a point where new irrigation systems are supplying increased amounts of water to the countryside, it also becomes possible to use substantial amounts of chemical fertilizer if it is available. Of course, in large parts of China

Table 1.   Selected Small-scale Industry Statistics.

| | Cement | | Nitrogenous Chemical Fertilizer | |
	Small Plants	Modern Plants	Small Plants	Modern Plants
	(million metric tons)		(million metric tons)[a]	
1957	0	6.9	0	0.7
1960	3.0	9.0	0	1.7
1965	5.4	10.9	0.5	4.0
1970	11.5	15.1	3.4	4.5
1975	27.7	19.4	10.3	7.4

[a]measured in standard units of 20 percent nitrogen.
Source: U.S. Central Intelligence Agency, *China: Economic Indicators* (Washington, D.C.: Government Printing Office).

Table 2.   Fertilizer Costs and Prices *(yuan per ton).*

	Large Plants 1956–57	Small Plants 1975
Ammonium nitrate	125 (625)	200 (1000)
Production cost		
Price to producer	-----	260–265 (1300–1325)
Purchase price to user	310–316 (1550–1580)	300 (1500)
Ammonium bicarbonate		
Production cost	-----	130 (763.18)
Price to producer		180 (1058.52)

Figures in parentheses are yuan per ton of nitrogen content.
Source: Dwight Perkins, ed. *Rural Small-Scale Industry in the People's Republic of China* (Berkeley: University of California Press, 1977), p. 99.

there were more than ample supplies of water long before the rural construction efforts of the 1950s and 1960s, but such was not the case in north China and it was in north China, where for reasons of supply to be discussed, small-scale industry was often most economic.

Improved supplies of both water and chemical fertilizer lead, in turn, to increased agricultural yields. In some cases increased amounts of water and chemicals make possible rising yields within the existing cropping pattern. In other cases the availability of water and fertilizer may make possible a change in the cropping pattern to two or even three crops a year. Whichever the case, the increase in output leads to a rise in the demand for labor. Where double- and triple-cropping are involved, as they are in most parts of China, the rise in demand during peak periods can cause an acute shortage of labor even in a country with a rural population of 700 million farming only 100 million hectares. Often within a period of a few weeks, farmers must harvest and thresh one crop and then prepare the fields and transplant (in the case of rice) a second crop. So it is that, even in labor-intensive China, there is room for a considerable amount of mechanization. Mechanized threshing (already extensively practiced) and mechanized transplanting (still in the experimental stage) can save a great deal of labor time at just these key periods of peak demand. Mechanization, therefore, may raise yields by facilitating double cropping or, at a minimum, by reducing the strain of workdays that begin at dawn and, where lighting is available, may stretch well into the night. Even if mechanization reduces labor demand in nonpeak periods, there is no danger of outright unemployment for anyone. The production team organization guarantees a job to all its members so that a reduced workload amounts to genuine leisure, not enforced idleness.

Once the process of rural mechanization is begun, it is essential to have repair facilities located near where the machines are being used. If a commune tractor has to be sent off to Peking or Shanghai for repairs, the commune may not see that tractor again for months. Nor will the large farm machinery manufacturers in the cities have rural distributorships, complete with spare parts inventories and mechanics. Services of this kind are usually found only in advanced countries. Thus, when mechanization began in a major way in China, it was accompanied by a large effort to ensure that communes and brigades developed their own repair facilities. As these repair facilities became developed and able to handle major repairs, they more or less automatically acquired a manufacturing capability. For example, because spare parts inventories were generally not readily available, a local repair shop had to manufacture its own replacement parts. The ability to make parts then often evolved into a capacity to make the machine itself. In some more advanced rural areas in China today, one sees small plants using a limited number of purchased lathes plus an allocation of steel to make more lathes'for their own use. These new lathes are then used to manufacture a wide range of types of farm machinery.

The main point of this presentation of a simplified but basically accurate picture of the sources of demand for the products of rural industry is to show how intimately tied this demand was to an overall effort to raise farm output and incomes. Cement, chemical fertilizers, and farm machinery formed the core of the rural small-scale industry program. No one of these industries is likely to have developed in the absence of a vigorous general rural development program.

ISSUES INVOLVING SUPPLY

Demand alone would not have been enough to bring rural industries into being, however. Demand has grown in other less developed countries without generating a small-scale industry program—let alone one emphasizing farm machinery, chemical fertilizer, and cement.

A large, poor country does provide industries located in rural areas with certain built-in advantages or sources of protection from larger more modern plants in urban areas. Poor distribution systems that make it difficult for urban plants to get their product to rural users on a timely basis are one part of the story. Chemical fertilizer that arrives a month after the correct time for its application will have a limited impact on productivity. Rural construction activities cannot simply stop for a month or two while their allocation of cement is awaited. In advanced industrial nations with many competitive firms in the same field and where the marketing-distribution system has been refined to a high order, unexpected delays of a month or two occur but are not common. In most less-developed countries they are the rule. In China, the normal problems of a less-developed country are compounded by the rigidities of the system of central planning and physical allocation of key materials that the Chinese adopted from the Russians.

Closely related to the weak distribution system is the problem of high transport costs. China does have a fairly well developed system of railroads and, through commune labor mobilization, a large number of rural roads have been built. But China still has few mechanized vehicles to run on those roads. Donkey-drawn and even human-drawn carts are still the most common ways of getting crops to market, although increasing numbers of hand tractors and a few trucks are now available for commune use. With vehicles in such short supply, transport costs are extremely high. In one county in Honan province, for example, the cost of transporting coal a mere twenty-five miles raised the price of that coal by 50 percent. Railroad and water transport costs, of course, were not so high, but much of rural China, particularly in the north, had no ready access to either railroads or easily navigable waterways.

High transport costs have had a particularly strong effect on the location and scale of rural cement and chemical fertilizer plants. Cement, for example, involves the transport of bulky (low per-ton cost) items such as limestone and coal to the plant and then the shipment of cement itself to users. Clearly transport costs could increase the price of cement to the final user by severalfold over what that user would have to pay if located next door to a cement plant that in turn had nearby sources of limestone and fuel. Under such circumstances, rural users located hundreds of miles from a major plant would probably not be able to afford to purchase any cement at all.

The solution to this transport cost-distribution problem in China was to locate small plants close to the rural users. Even if operating costs for such plants were substantially higher than those of large-scale plants, the elimination of transport would still make costs to the user lower. As the rural distribution-transportation system improves, of course, this source of protection for rural industry will gradually be reduced. For the smallest and least efficient rural factories the end of protection will probably mean the end of their existence. But for the somewhat larger and more efficient rural plants, such need not be the case. As rural industrial skills improve and as China learns more about small-scale technologies most suitable to the countryside, costs will decrease. In fact, such a trend toward larger, but still relatively small-scale, rural plants is already apparent in China and, although the evidence is limited, the costs of these plants do appear to be coming down.

As transport-distribution costs decrease, rural industrialization might actually gain in strength. Today relatively few rural plants produce for a market larger than that of the county in which they are located. In part this limited market is a result of the Chinese system of planning. If an enterprise sells a significant portion of its product beyond the borders of the county, then county planners will lose control over the enterprise. Control will pass, instead, to a higher-level administrative unit (district or province) that has authority over all the areas in which the enterprise's products are sold. Since one of the appeals of rural industries to local planners is that the local people control what is produced, there is resistance to gearing up to produce for a larger market. But even if such resistance did not exist, high transport costs would produce much the same result. Thus, as transport costs decrease, the possibility of producing for a larger market will become a reality for rural as well as urban plants. Using urban plants as subcontractors for larger urban enterprises will also become possible. Today such subcontracting exists but mainly in rural areas within one or two hour's train ride of a major city.

High transport costs are not rural small-scale industries' only means of overcoming the advantages in terms of efficiency of large-scale urban plants. Labor and material costs also tend to be lower in the countryside. And if these costs were calculated at their social cost to the nation rather than what factories actually paid for labor and materials, rural costs would be even lower.

In China, factory workers in rural plants receive slightly less than urban workers doing comparable jobs. In factories visited by the American Rural Small-Scale Industry Delegation, for example, the average wage in eighteen of the rural plants visited was 41 yuan per month whereas in thirteen of the urban factories for which data were given, the average wage was 61 yuan per month or nearly 50 percent higher. Some of this difference may reflect a higher skill mix among urban workers, but it is unlikely that that is the main source of the difference. Rural workers are paid less in part because the government does not want their incomes to differ too much from those of the farmers who are their neighbors. Urban wages also reflect the higher cost of urban living. For society as a whole, of course, the higher costs of urban living—reflecting the high cost of transporting fuel and food to the city, of building sewage systems, and the like—are real costs to the nation.

In another way, money wages may exaggerate the real cost of labor to both rural and urban plants. From a social point of view, the issue is what that labor would be doing if it were not in the factory— that is, what is the opportunity cost of removing labor from agriculture. Except in peak seasons, the marginal product of agricultural labor is usually very low. Removing labor on an off-season basis, therefore, involves little real cost to society. Many rural plants do in fact encourage workers to help out on the farm in peak seasons although this is not possible in continuous process operations nor is it common in the larger rural plants in general. To the extent that rural factories can operate on a seasonal basis, however, the real labor cost is clearly lower than the rural money wage and far lower than both the money and social cost of urban labor.

A similar kind of opportunity-cost argument can be used in the case of some of the material inputs into rural small-scale industry. China, particularly north China, is endowed with numerous widespread outcroppings of such key inputs as coal and limestone. Such small outcroppings of coal, for example, have been a source of fuel for home heating in China since before the days of Marco Polo. In recent years these same croppings have been the main source of fuel and feedstock for the local chemical fertilizer and cement plants. Many of these small mines and quarries could not be exploited economically if their output had to be shipped long distances. But the opportunity cost of using local labor to dig out and transport this coal and limestone to nearby plants is very low. The labor involved does not have to be skilled for the most part, nor do these mines and quarries use much in the way of machinery. Much of the labor, in fact, could be supplied during the off-season when the alternative would be to work on low-productivity rural construction projects.

Thus in a capital-poor and land-poor rural society such as that of China where transport costs are high, rural industry has built-in cost advantages as long as the market for its products is not too far away. As economic development proceeds, not only will transport costs decrease, but the opportunity-cost of labor will rise as productivity in agriculture itself increases. If small-scale techniques improve at a rapid enough pace, these declining cost advantages need not spell the end of rural industries although it is the larger end of the small-scale spectrum that is most likely to survive. Even the survival of larger rural plants will depend on a research and development effort focusing on making such plants more efficient.

THE ROLE OF THE CENTER

Successful research and development, of course, will not be carried out by local authorities acting alone. The laboratories, scientists, and other technical support will have to be developed at the provincial and central levels. Local tinkerers can make minor improvements but it makes no economic sense to have thousands of counties and communes duplicating one another's research efforts of a more fundamental kind. China today already has a variety of research institutes at the provincial and central level working on the problems of small-scale industry. With the demise of the "gang of four," the expectation is that the emphasis on the human and material resources given to these research institutes will be greatly increased along with the general effort to upgrade science and technology.

It is not just in the area of research and development, however, that rural small-scale industry requires the support of the center.

Even in a nation such as China with a vigorous rural development program providing local demand and mobilizing local labor resources, the county and commune authorities cannot perform all the necessary tasks alone. In some very poor areas the state may have to help out with injections of outside capital. Capital, however, is not a major problem in better-off areas. Local planners do not need to induce individual peasants to invest their small savings in local industries. All these planners have to do is to decide to use some portion of the commune investment fund or of profits of other local industries.

Where central support is crucial is in training key personnel and in supplying key inputs and equipment not available locally. Steel, for example, is sometimes produced from scrap in local plants, albeit ones far larger than those that characterized the backyard campaign, but most rural areas receive their steel allocations from higher-level plants. Certain equipment in chemical fertilizer factories is also beyond the production skills of local plants, and the same is true but to a much lesser degree for cement plants. Motors for hand tractors are produced in large enterprises in Peking and elsewhere, and the list can be extended. Still, what impresses the outside visitor to rural China is how much of the output of the small plants is produced from local inputs.

Small plants, of course, would never have succeeded if each had had to learn on its own or from books what was required. Books explaining small-industry techniques do receive wide circulation, but much of the initial technical expertise in local plants is acquired by sending workers and technical personnel to other small plants or to large-scale plants in the cities that produce similar products. The absence of proprietary rights over the technology is probably one reason one plant is so willing to help set up another. More important is the fact that training others is something that all enterprises are expected to do, and they receive rewards or criticism for successes and failures in this area. A common sight in small rural enterprises is to see two workers to a lathe where one would suffice. The reason is not an economically inefficient desire to maximize employment, but instead a desire to train new lathe operators either for an expansion of their own plant or to staff a new plant. On the walls of some of the more efficient plants that serve as models for the others, a visitor sees banners from other factories thanking the model for its help in getting them started or raising their technical level.

How difficult is it for the center to ensure that local enterprises receive the inputs and help that they need? Certainly the supply of help would not have been available without a conscious effort on the part of the center. Steel is in short supply in China, and the managers of large plants have far more clout when it comes to obtaining supplies than managers of small plants. There is no free market for steel in China, so rural enterprises cannot simply purchase what they need even if they have sufficient funds. Nor would small factories be in a position to bid away skilled personnel from larger plants or to insist that the larger plants train their workers. Someone with authority over the larger enterprises had to provide those enterprises with the incentive to undertake these tasks. In China's command economy, the incentives came from plan targets or orders that managers knew would be enforced with sanctions. The effort required on the part of the center was not great given China's highly authoritarian command structure, but an effort did have to be made. The effort involved was also reduced by the fact that Chinese industry had grown to a considerable size by the time of the resurgence of the effort to encourage small-scale industry. At no time did supplies of steel, electricity, machinery, and technicians required by rural industry represent more than a small fraction of the total available supplies of these items. Put differently, rural small-scale industrialization depended in a fundamental way on the prior and continuing successful development of urban large-scale industry. The two sectors were more complements than substitutes.

EXTERNAL BENEFITS
OF RURAL INDUSTRY

Thus far this discussion has been confined primarily to an analysis of the direct economic benefits and costs of rural industry within the context of a national commitment to promote rural development. This commitment to rural development is, of course, a social as well as an economic objective. A commitment to rural industrialization has benefits for the quality of rural life that extend beyond the fact that such industries help raise rural per capita income.

Perhaps most important among these benefits is the impact of rural industry on the spread of skills and knowledge of modern technology. In part this spread of technique is simply another economic gain, but it represents a gain not confined to the rural industries themselves. The presence of factories and technicians in or near the villages leads to an enhanced knowledge of how modern technology works on the part of family and neighbors as well as the factory employees themselves. But familiarity with modern technology has implications beyond the economic sphere. Traditionally rural societies including China's have been characterized by attitudes of fatalism on the one hand and a belief in the almost magical properties of modern technology on the other. The rural presence of workers building and using modern machines tends to break down both attitudes. Machines are demystified and in the process people learn how they too can use modern technique to control their own lives.

Another benefit of rural industrialization, in the eyes of most people at least, is the impact it has on changing the traditional role of women. Spinning of yarn and milling of grain, for example, have usually been women's jobs in China and onerous jobs at that. Mechanization of both of these processes either in small rural plants or large urban factories has thus made women's life easier. Rural industries have also provided employment opportunities for women outside the home. The management and technical jobs in China's rural industry are still dominated by men, but women are found in much greater numbers at lower-level jobs. In twenty-two small factories visited by the American Rural Small-Scale Industry Delegation, for example, an average of 25 percent of all employees were women. Alternative employment for women, among other things, means that for some women at least it is not foreordained that they will move from a father to a husband and mother-in-law-dominated farm household. They have more choice in the matter than when no such alternatives were available.

A third noneconomic benefit of rural industrialization from the point of view of China's leaders is that it strengthens the role of the commune- and county-level administrations. In the aftermath of the failures of the Great Leap Forward, the commune staff lost most of its reasons for existence. Farm management, for the most part, was handled by the production team. Except for the milling of grain and low-level machinery repair, however, the team and even the brigade (made up of a number of teams) were units too small to take responsibility for the operation of rural industries. Thus rural industrialization gave commune authorities both a role and a source of income. For a leadership concerned with slippage back toward private peasant agriculture, the strengthening of the more socialist commune organization was a definite plus.

Surprising to some is the fact that reduction of rural unemployment has not been a major objective of the rural small-scale industrialization program. In a literal sense, to be sure, there is no unemployment in rural China. Virtually everyone not employed elsewhere is a member of a production team and is guaranteed employment in that team as well as a share of the team's income. But employment in agriculture, particularly in the off-season, may mean little more than the arduous and unskilled effort of digging and carrying rock and dirt. Thus employment in rural industry would represent a marked improvement for many in the quality of their workday. To Chinese leaders, however, the main task of both rural labor and rural industry is to raise agricultural output. If the creation of rural industry draws workers away from farming, the task of raising farm output is made just that much more difficult. Hence, wherever labor-saving devices can be installed cheaply, they are strongly encouraged so as to allow farm activities to retain as much labor as possible.

Despite the lack of emphasis on the employment possibilities of rural industrialization prior to 1979, the underlying general purpose of the small-scale industry program can be seen as an effort to avoid the extreme of the dual economy that is characteristic of so many developing countries. Rural industry is a substantial indication of the government's desire to see that rural as well as urban incomes rise. The rural sector is itself expected to participate actively in its own modernization. Modern technology is not simply something to be handed down from above by urban planners and urban factors. The rural areas are to supply a good deal of that modern technology through their own efforts. Dualism in China still exists, to be sure. The urban areas are still the center of the economic modernization effort, and many rural areas lag far behind. But the lag is smaller than it would be in the absence of the widespread development of rural small-scale industry.

ARE THE LESSONS OF CHINA TRANSFERABLE?

The question uppermost in the minds of most non-Chinese observers of the rural small-scale industry program is whether or not the Chinese experience is transferrable to other developing countries. The major problem with any discussion of transferability, of course, is that it is difficult to discuss the issue in general terms. The central question is "transfer to whom?" There is little point in talking about a major small-scale industry program for a nation with a weak urban-centered government and no general rural development effort to speak of. All that can be attempted here, therefore, is to suggest which of the elements described in the foregoing were crucial to the success of the effort in China. It is a task for others to decide whether comparably favorable circumstances exist elsewhere.

Several of the elements central to the development of rural small-scale industry in China are not, on the face of it at least, unique to that country. High-cost transport and a weak distribution system clearly fall into this category. An industrialization effort capable of supplying both equipment and training to rural industry also exists elsewhere. Not all industrialization in the 1950s involved the rapid development of machine tools and other producer-goods industries. An industrialization program centered on textiles and electronics, although desirable on many other grounds, might not be as useful a support to rural producers of cement, chemical fertilizers, and farm machinery.

Strong central government support and an existing major rural development effort at all levels were both important sources of assistance to rural industry. Clearly a great many less-developed countries lack both, but China is not alone in possessing both. For nations lacking these "preconditions," all that really can be said is that their rural development prospects are dim and that rural industrialization will not provide an easy means of escape. Rural industry, after all, is not a substitute for other rural development programs. It can only be a part, sometimes a crucial part, of a more general effort.

Central to any appraisal of the transferability of China's rural industrialization effort is the degree to which that effort depended on the prior existence of communes and the socialization of industry. If both were crucial, then there is little point in talking about the relevance of China's experience for nonsocialist countries.

The commune form of organization did play a positive role in support of rural industry. It was the commune that organized the vast amounts of rural labor that built the rural construction projects that provided so much of the demand for local cement plants. And it was the commune that could readily extract a percentage of farm income to invest in rural factories. Consolidated large fields also contributed to the ease with which machinery could be introduced. But fields have been consolidated within the context of individual peasant agriculture, and methods of sharing the larger and more expensive machines have been devised. The method of extracting funds for investment will differ in the absence of communes, but methods consistent with private agriculture can certainly be devised if the power and will exist. Nor was the construction of small- and large-scale irrigation systems, roads, and the like invented in China after 1955. In short, the commune-facilitated efforts had a direct bearing on the success of rural industrialization, but there is no reason to think that only the commune could facilitate such efforts.

Much the same kind of argument can be made about the relation between state ownership of industry and the rural small-scale effort. Some, perhaps many, rural industries ran at a loss for a number of years before becoming profitable. But there are many ways of subsidizing private firms to carry them over the difficult development period. Enterprises that run at a loss (in social accounting terms) for a long period or permanently should generally be abandoned whether in China or elsewhere. Similarly the absence of proprietary rights eased the problem of transfer of technology. But nonsocialist countries can still have public research and development efforts or can pay the private sector adequately for discoveries made there and then distribute the results to others.

Finally, there is the question of the degree to which China's rural industrialization effort depended on a particular kind of political system. Certainly such an effort would have been impossible if county and commune cadres were automatons obeying the orders of higher authority. Rural industries had to be adjusted to local conditions in a variety of ways that made centralized direction from Peking or even a provincial capital inefficient. In fact one of the great appeals of rural industrialization to local cadres was that such industries gave them a much higher degree of flexibility in designing rural development activities in a wide variety of areas. A local project requiring cement or machinery no longer required high-level approval because most of the required inputs could be made available from local sources.

The existence of decentralized authority in certain areas, however, should not be confused with democratization in the Western sense of the word. The local population plays a role in deciding how to implement central objectives, but the great mass of the people have little influence on the objectives themselves. The national commitment to rural development, therefore, is a commitment of the leaders of the Chinese Communist Party, and not the result of a national plebiscite or some other form of expression of the popular will. A democratic country with free elections (in the sense the word is used in the West), however, could conceivably produce the same results. After all, the great majority of people in most less-developed countries live in the rural areas and have a vested interest in a greater rural development effort.

A more realistic issue, given the rarity of democratic regimes in the developing world, is whether the various forms of authoritarian government that do exist are compatible with a broad rural support program. Certainly an authoritarian government dominated by an urban elite interested primarily in its own perquisites will not promote such an effort. A government dominated by the interests of rural landlords also may not be capable of initiating a major rural development program. It is difficult to mobilize rural people for a broad-based effort when most of the benefits go to a small minority who own most of the land. Rural industrialization must be a part of such a broader effort. But all we are saying here is that a government that does not really want rural development, as shown by an unwillingness to allocate scarce resources to such an effort, will not succeed with rural industrialization. The Chinese Communist Party was able to make a commitment to rural development in large part because it was a party that was brought to power by the rural poor and whose core base of support still depended to a degree on those rural poor. Perhaps the central question concerning the transferability of China's rural industrialization experience, therefore, is whether governments that have not come to power as a result of a revolution of the rural poor can make a similar commitment.

If we leave aside the question of whether other kinds of governments can make a commitment to rural development, then there is little in the Chinese experience with rural small-scale industry that is, by its very nature, unique to China or to collectivized agriculture and state-owned industry. At the same time China possessed a considerable variety of institutions and policies that in combination were central to the success of rural industrialization. Nations with very different institutions and policies can, in principle, design substitutes for what the Chinese did that are more appropriate to their own societies, but the task is not a simple one. Successful rural industrialization like that which is occurring in China requires a broad commitment to rural development and a willingness to make major changes in policies and in institutions to make such a broad commitment a reality.

✳ *Chapter Ten*

On Being Rich and Being Poor: Technology and Productivity

Kenneth E. Boulding

If the wealth of nations, as Adam Smith suggested, is to be measured by their per capita real product, then differences in the wealth of different nations must depend on differences in this variable. The key to the differences, and therefore to the distribution of income, lies in the determinants of real per capita product in some kind of value terms in different nations or groups.

I devised a formula[1] more than twenty years ago that seems to me to incorporate the main determinants of per capita real product, and, although I have found practically nobody else who shares my enthusiasm for this formula, it can do little harm to repeat it in a brief and slightly revised form. If for any individual or group, Y is the total product of economic goods, P is the total population, then if h is the number of hours in a year, Ph represents the total human resource, that is, the total number of person-hours that the unit in question has to spend in the course of a year. We can divide Ph into three parts: D, hours spent during a year producing for domestic use and consumption; E, hours spent for export, that is, exchange with other groups; U, hours spent not producing economic goods. U is a slightly tricky concept. It could be argued, for instance, that sleep, by restoring the value of a person, produces an economic good offsetting human depreciation. The same could be said of time spent eating, time spent in sex, childrearing, and so on. These activities tend to consume a fairly constant number of hours, so that variations in U mainly consist of unemployment and leisure, time that might be spent producing economic goods but in fact is not so spent either voluntarily or involuntarily. Suppose that p_d represents the productivity of time spent in producing for domestic uses; p_e, the productivity of time spent in producing for export; T, the terms of trade of exports, that is, how much imports of domestically used and consumed goods are received per unit of export goods; and G, which can be either positive or negative, the grants of economic goods, that is, one-way transfers (such things as welfare, foreign aid, tribute, gifts, and so on). We then have

$$Y = Dp_d + Ep_e T + G \tag{1}$$

$$Ph = D + E + U \tag{2}$$

and from these we get, if the per capita real product per hour is y,

$$y = \frac{Y}{Ph} = \frac{Dp_d + Ep_e T + G}{D + E + U} \tag{3}$$

a slightly modified version of my original formula. This throws a good deal of light on the forces that lead to a rise in y, that is, an increase of riches. Of course, h is an arbitrary constant defining the time unit, which we can neglect. We then trace several processes by which y can be increased.

1. If U is diminished, $D + E$ will increase, and as p_d, p_e, and T are positive, y will increase. It is observed in recovery from a depression, with declining unemployment, in the changing composition of the population producing an increase in the labor force, and so on. We have to be careful about interpreting these changes, particularly where not all significant products get into the official statistics. Thus, a shift of children out of the labor force into education may produce a decline in y, but it may also produce an improvement in the welfare of the society because the product of education in human capital is larger than the loss suffered from the diminution in immediate goods. Similarly, a shift from home industry to factories often results in a spurious expansion of y, because the loss in home products is not recorded in the statistics. We always, however, have to be on the lookout for indicator failures. Indicators are always evidence rather than truth and always have to be interpreted.

2. Even if $D + E$ is constant, if one increases at the expense of the other, then an increase in y may take place. Thus, if $p_e T$ is greater than p_d, a shift from D into E, that is, from domestic production into production for export, will increase per capita real product. If p_d is greater than $p_e T$, a shift from export into domestic production will increase real product. Product is maximized, other variables

being constant, when the division between D and E is such that p_d is equal to $p_e T$. The actual dynamics of this system may be fairly complex. However, it is not unreasonable to suppose that as we shift out of D into E under the stimulus of $p_e T$ being greater than p_d, p_d will increase as the least productive domestic activities will be given up first, whereas p_e may decrease as we move into less-productive export activities and T will also decrease as an increase in exports tends to lower their price relative to imports. It is not unreasonable to suppose, therefore, that there is an invisible hand at work here— that if, for instance, export industries are "too small" relative to domestic industries, export industries will be more profitable and domestic resources will move from domestic into export industries, which will lower the profitability of domestic industries until the movement ceases. The possibility remains open, however, that the point at which movement between D and E ceases is not necessarily the point at which p_d is equal to $p_e T$. If this is so, a case for intervention may appear.

3. It is clear that any increase in T, that is, an improvement in the terms of trade, other variables being constant, will increase real per capita product. A decrease in T will diminish per capita real product. There is certainly no harm in having good terms of trade. T, however, is unlikely either to increase or to decrease for very long. Historically, it tends to oscillate, rising in one period, falling in the next, for the very forces that make it rise or fall tend to limit the movement. Thus, coffee-producing countries have been benefiting substantially by the high price of coffee in recent years, but that very high price encourages production in a wide variety of places, and in the absence of monopoly, seals its own doom. Similarly, the low price of copper is certainly bad for the copper-exporting countries, but this will discourage copper production and eventually raise the price of copper. The terms of trade, therefore, are essentially short-term. They cannot account for any sustained rise in per capita real product. There has been a great deal of debate as to whether terms of trade are not permanently unfavorable for certain products, such as agriculture and raw materials. The evidence for this is perhaps stronger in the case of agriculture than it is for other raw materials. There seems to be a good long-run reason for it: in the course of economic development and particularly as agricultural productivity increases, the low-income elasticity of demand for agricultural products results in a constant transfer of people out of agriculture, at the same time that increased productivity enables agriculture to feed the rest of the population with a smaller and smaller labor force. This transfer of people out of agriculture, however, seldom takes place fast enough

to correct the poor terms of trade. Agriculture may be "squeezed" for a long time in the process of getting people out of it. It is probably less true of other raw materials, which often tend to be incorporated in products that have high-income elasticities. There has been some decline in the proportion of the labor force engaged, for instance, in mining, but much less than is the case with agriculture, where the decline is spectacular—in the United States, for instance, from about 90 percent two hundred years ago to under 4 percent today.

4. A positive G—grants of economic goods—increases y, and a negative G diminishes it. Here again, however, we are very unlikely to have a sustained rise in G. There has been a slow rise in grants to the poor from the national state, but if we are not careful these can easily be offset by adverse movements in the terms of trade of the poor. Much depends here on the nature of the grants. On the whole, the grants element in riches has been relatively small, except within the family. With increased skill and sophistication in the tax system, this element may become substantially larger. A redistributive income tax, for instance, with both positive and negative components, could markedly affect the distribution of income. On a world scale, G is very small, whether positive or negative. Exploitation is a grant that is regarded as illegitimate by at least somebody. In certain times and places this is noticeable but is again not capable of sustained expansion. Monopoly profits are perhaps the best example, plus tribute obtained by war or conquest, but outside of exceptional cases these tend to be quite small.

5. This leaves us with the productivity elements, p_d and p_e, as the only factors in the formula capable of sustained growth over long periods. Even here, the capacity for sustained growth in productivity varies greatly from industry to industry. It has been greatest perhaps in food production. Thus, the development of agriculture, even though it had some unfortunate side effects, undoubtedly increased the productivity of food production over that of hunting and gathering societies: a society could produce all the food that it needed using a certain fraction—perhaps 80 percent of the population in classical civilization—while releasing the rest to build cities, fight wars, practice the arts and religion, and indulge in all the other delights of civilization. The rise of science has permitted an even more spectacular rise in agricultural productivity, to the point where now in developed societies 2 to 4 percent of the labor force (or perhaps if we include the other energy and materials inputs, up to 10 percent) can produce all the food that society needs, releasing 90 percent or more for other things. In industry, likewise, there has been steady, though somewhat slower, increase in overall productivity, reflected

perhaps more in the increased output of industry than in a change in the proportion in the labor force devoted to industry. This proportion has been surprisingly stable in the rich countries, at least over the last few decades.

The expanding industries are the tertiary ones—government, education, the arts, science, research, information and knowledge industries, and so on. Here increase in productivity is more difficult to achieve, although it is possible. It is not impossible to increase the productivity of simple knowledge systems, like accounting or airline reservations. Even if there is no way of increasing the productivity of a string quartet, recorded music has enormously increased the number of people who can hear music. More people listen to Mozart today in a single hour than listened to him in his whole lifetime. Similarly, color printing has created a "museum without walls" and vastly expanded the artistic experience of the human race in quantity, whatever we might think about quality. On the other side of this coin, technological change in defense has almost certainly diminished everybody's security and has increased the real cost of war to the point where it has become almost unbearable. The productivity of civilian government is extraordinarily difficult to measure. One has an uneasy suspicion that it has been diminishing by the sheer role of Parkinson's law—that bureaucracies feed on their internal information processes without much increase in actual output. In education, it is doubtful whether productivity has increased very much since the days of Plato. The classroom was a great technical invention, which certainly increased the productivity of the learning industry over that of the mother's knee. Since then, however, apart from a few gadgets like microfilms and overhead projectors, the education industry has remained very much what it was in Plato's day, a teacher getting up in front of a class and talking to it.

One of the great puzzles in world history is why increases of productivity have taken place in some societies and not in others. There seem to have been two such major increases in human history. One stemmed from the invention of agriculture and the subsequent development of cities and civilization, the other from the rise of science, which had an enormous impact on human know-how and productivity, and is still very far from having exhausted its potential. Both changes led to substantial shifts in human population, with those populations that participated in the change expanding both numerically and geographically at the expense of those populations that did not.

One of the still puzzling questions in human history is the relative importance of a cultural spread of new technologies, that is, the adoption of increased productivity techniques by people who have not used them before but who imitated others among whom perhaps they had originated, in comparison with the "demic" spread of the population in which new know-how and technology arise, geographically displacing those populations that do not adopt the new techniques. A number of cases of the latter have occurred. For instance, genetic evidence suggests that the spread of agriculture from the Middle East into Europe was demic rather than cultural,[2] that is, it took place through successive invasions of people who adopted agriculture, displaced the earlier peoples, and pushed them to the north and west. A similar phenomenon was the invasion of the Americas by Europeans after Columbus. What might be called an "eoscientific technology," involving guns, ships, sails, horses, and so on, as well as agriculture, displaced the native populations, especially of North America (to some extent, also, in South America) in a process remarkably similar to what seems to have gone on several thousand years earlier in Europe. On a smaller scale, the displacement of the Ainu by the Japanese as they moved from their origins in the southern part of the Japanese islands northward into Kansu and eventually to Hokkaido represented the same demic expansion of a technologically superior culture. Similarly, the Bantu expanded from central Africa into southern Africa following their adoption of agriculture; here they met the Europeans, especially the Dutch coming up from the south, both of them displacing the original habitants, the Hottentots, with a technologically superior culture. Australia and New Zealand are another example of demic expansion, with Europeans displacing the original inhabitants. The Aryan expansion into India about three thousand years ago, pushing the original Dravidians to the south, was another example.

Just why a displaced culture does not adopt the technology of the culture that is displacing it remains puzzling. Sometimes indeed there is cultural rather than demic expansion. Japan is a good example. When threatened with the "black ships" of Admiral Perry and the pressures of a technologically superior culture from North America and Europe, the Japanese adopted the culture with great rapidity, and indeed in many ways became more successful than its originators. China is an interesting case, somewhat in reverse, where all the conquerors adopted the technically superior Chinese culture and the Chinese seem to have taken very little from their conquerors.

Further inquiry into the nature of productivity increase demands a more adequate theory of production than the one economics has provided for the last two hundred years. Economists tend to adopt what might be called a cookbook theory of production, in which the product emerges from mixing together in various proportions land, labor, and capital. Marxism goes even further than this with the one-ingredient recipe, in which labor is the only factor, capital simply being embodied labor of an earlier period. These theories seem to me to be unsatisfactory. I have argued that production essentially in an evolutionary framework is the process by which the genotype produces the phenotype, as the egg produces the chicken or a blueprint a house. All production originates from some kind of information structure possessing "know-how," that is, a program capable of guiding a process by which energy is directed toward the selection, transportation, and transformation of materials into the improbable shapes and structures of the product. Evolution is primarily a process in genetic structure, that is, know-how. It can, however, be limited by the absence of the requisite energy and materials. Under favorable circumstances, however, an increase in know-how will push back the energy and materials limitations, which will increase the "niche" of the product, or phenotype.

We see this happening many times in human history. The discovery of fire, for instance, undoubtedly expanded the human niche substantially at the expense of competing species that did not have it. Fire represented an increase of the energy input into the social system beyond the muscular energy of human beings themselves. Wood was in a sense the first fossil fuel, representing as it does fossil sunshine. Eventually this led to a whole new range of materials— pottery and metals—as humans were able to transform materials at higher temperatures. At all stages there is a complex interaction between new sources of energy and new materials. Agriculture, for instance, represented increased productivity in the use of solar energy, not only in plants but also in the domestication of animals that seems to have gone along with it, and again expanded the energy input of the system beyond that of the human muscles. Metallurgy, from gold and silver to copper, bronze, iron, and steel, now to aluminum and titanium, again expanded the materials input of social systems and helped to produce corresponding changes in the social structure. The rise of cities, for instance, seems to have been closely related to the development of metal weaponry that could subdue the agriculturalist and extract his surplus for the feeding of conquerors and kings, soldiers, builders, artisans, artists, musicians, and priests. The rise of science, for some mysterious reason taking place in Europe beginning some five hundred years ago, represented another profound mutation in the evolutionary process itself, with the development of much more accurate "know-what" as reflected, for instance, in the chemical elements as against the hopeless aggregates of the alchemists—earth, air, fire, and water—as well as in the physical principles that have produced modern engineering and space travel.

We thus see productivity as a result of the complex interaction of the different species that constitute what might be called the "economic ecosystem." First of all we have know-what, as embodied in academic and folk knowledge in both the sciences and the humanities. This has expanded enormously, especially in the last two hundred years, as a result of the development of the small subculture of science with the high value it placed on curiosity and veracity, communication of results, and public criticism of them. Know-what increases as a result of two processes: mutation in human images, that is, the development of new theories, new hypotheses, and postulates—and selection, which is the process of testing in its many varieties—experimental, observational, statistical, and so on. Direction in evolution occurs when there is asymmetry in the selection process. In the case of human images, the asymmetry reflects itself in the fact that erroneous images are more likely to be changed in a testing culture than are true ones. Science may do little more than substitute ignorance for error, but even this is a great advance over the persistence of error.

The impact of this accumulating know-what was very slow at first. Its major impact on production and economic life indeed took place only after 1860, with the development first of the chemical industry, and then the electrical industry, the nuclear industry, and various biological industries. The great age of change was from about 1860 to 1920, which saw the introduction of chemicals, electric power, automobiles, the internal combustion engine, airplanes, steel-frame buildings, telephone and telegraph. Photography and the steam engine were a little earlier, but not really very much. In my lifetime I have seen television, computers, jet airplanes, but these represent a much smaller impact on human life than did electricity, the automobile, and the telephone. Social inventions likewise are important. For instance, the rise of large-scale organizations after 1870 had something to do with the telephone, the typewriter, and the secretary, but it also represents some social inventions in organizational structure that eventually produced General Motors, the Pentagon, and the Soviet Union, all of which would have been inconceivable in 1860.

Neither know-what nor know-how, neither science nor technology, are merely additive. They are strongly interactive. Changes are not simply piled on top of each other. Each change opens up the possibility of many other changes, in a highly complex network. The developments of the eighteenth century, for instance, have been much exaggerated, for the world was not much richer in 1800 than it was in 1700, nor was its technology drastically different. There was, however, improved agriculture and the development of the turnip, clover, four-course rotation, and "horse hoeing husbandry" of the early part of the eighteenth century, leading to a release of population from agriculture and an improvement in nutrition, which unquestionably led to a decline in the infant deathrate and an expansion of population. In England and Northern Europe this provided an opportunity for new manufacturing towns, for the improvement in textiles, in themselves relatively minor, or what was more important, for improvements in metallurgy. A key invention of the modern world was Abraham Darby's discovery of how to smelt iron from coal, about 1740. This introduced a whole new energy resource into the economy, allowing a further expansion of society. Chemistry led not only to analine dyes, but to phosphates and artificial fertilizers, which expanded agricultural production again, releasing more people for industry and services and leading, for instance, to the Bessemer process for steel, which in a sense built the skyscrapers, with innumerable smaller things aiding and abetting the larger process. The history of the "great change"—the extraordinary period between 1860 and 1920—remains to be written. Popular images are still dominated by the illusion that there was an industrial revolution in the eighteenth century that somehow created the modern world. The eighteenth century was simply the tag end of the Middle Ages, or at least what might be called the "eoscientific period," on which, however, the expansion of Europe was originally based.

One of the real mysteries of the world is why scientific technology penetrated so slowly into the tropics, and even into China, which had led the world in technology for almost 2000 years before the 1500s or 1600s. Thus, it may indeed be that the expansion of the European peoples into the temperate zone in North America, the southern part of South America, South Africa, Australia, New Zealand, and Siberia, was demic rather than cultural. It may be easier to invade than to convert. The great exception to this, of course, is Japan, though the history of Japan after 1868 is not so different from Sweden. It may only have been disease that prevented the demic expansion of Europe into the tropics, and, if so, it was very effective. Populations of European origin in the tropics, even in the Americas, where they are largest, have remained quite small.

One should add that there is nothing "biogenetic" about development. All human races have very similar biogenetic capacities. Development is essentially what I have called "noogenetic," that is, it rests on the development of the potential for knowledge and know-how in the biogenetically produced human brain, which is capable of being structured by a learning process, structures of which can then be transmitted from one generation to the next. The transmission of these structures, however, seems to be easier within homogeneous cultures than it is across diverse cultures, simply because of the degree of contact of one generation with the next, particularly where the learning process occurs primarily in the family. Cross-cultural learning is often extremely difficult. This is probably why earlier expansions of technology were mainly demic rather than cultural. Now, however, the situation is very different. Science and science-based technology are transmitted not mainly within the family, but within the formal educational institution, the apprenticeship system, the corporation, and other organizations. This means that the obstacles to cultural transmission of knowledge and technology have been greatly diminished. Japan, perhaps, was the first major example of this. In the Meiji Restoration the young Japanese learned their science and engineering not from their fathers but from European teachers, and then were able to transmit this accordingly to future generations of Japanese. There is no reason why this process cannot now be repeated all over the world, and indeed cultural development is proceeding with great rapidity. Nairobi, Delhi, Singapore, Manila, Belem—all have universities and these institutions are part of a worldwide system of transmission of publicly available knowledge from one generation to the next. We learn our native language and our folk culture in the family and from our immediate peers. This leaves the world culturally divided, also culturally diverse and rich in evolutionary variety. What we learn in universities is a world culture, especially in the sciences—the periodic table appears on the walls of chemistry lecture rooms in every university in the world. There is no such thing as Chinese chemistry or Catholic chemistry, Communist chemistry or African chemistry, though there will be local variations when it comes to medicine, engineering, and construction, and still greater local variations in politics, economics, and family life.

The haunting and tragic problem is that of the impact of the "superculture," comprising universities, airports, automobiles, telephones, and bureaucracies, on the folk cultures which lie across its path. The tragic history of the American Indian and the Appalachian whites is now being repeated in Africa with the Ik, and perhaps the Zulus; in Russia, with the Uzbeks and perhaps the Tartars. When the steel vessel of the superculture presses against the clay pot of more traditional societies, the clay pot often shatters. The older society can neither adapt nor withstand the pressures of the new and it sometimes collapses into a terrible anomie, and the human suffering and wastage involved are a nightmare. One can see perhaps the Gandhian village movement in India, or even the cultural revolution in China, or the villagization policies of Nyerere in Tanzania as an attempt to stave off the devastating impacts of modernity. It is an irony indeed that one senses in the socialist countries a certain Victorian charm, an isolation from devastating impacts of the modern world and the superculture. But even in Edwardian Bucharest one finds the Intercontinental Hotel in the best neobrutal modern style. There are glass shoeboxes now in Moscow and no doubt they will shortly be in Peking. In London, Paris, Tokyo, Nairobi, even Philadelphia, one sees the superculture destroying an earlier, smaller-scale, more elegant society.

A great question for the next hundred years is: Will the triumph of the superculture be complete? Will the kimono go the way of the crinoline? Will the water buffalo be exterminated by the tractor and peasants everywhere turned into superpeasants on the Iowa scale, with agriculture reduced to 4 percent of the world population? Within a century will Kenya look like Colorado and Bengal like Florida? I suspect that the answer is almost certainly "No," because of the sheer number of people involved (in spite of the slowly declining rate of population increase in the tropics) plus the high probability that the energy and materials limitations that were pushed back very sharply in the nineteenth century by the discovery of oil, natural gas, and large bodies of very cheap ores will creep slowly toward us. We are not sure of this, for technology is notoriously unpredictable. We may have a dramatic breakthrough in fusion or in solar electricity, perhaps from photovoltaic cells. Electricity, at least, may turn out to have quite an elastic supply at a reasonable price. On the other hand, electricity is not the same as fuel. It will not run automobiles efficiently. The problem of energy storage becomes increasingly acute as we move from fuel to electricity and solar energy. High technology is unlikely to disappear and may well become even higher perhaps as we move out to space colonies, but it is unlikely to displace the simpler technologies and indeed may even provide niches within which these technologies will survive or even flourish.

On the other hand, one is conscious of the gap between the richest and the poorest nations of the human race with respect to both the know-what structure of science and the know-how structure of technology. Science has always been a temperate zone subculture. It has devoted itself very little to tropical ecosystems. We do not know enough about their stability under pressures of change. Furthermore, there is a good deal of evidence that tropical ecosystems in fact are rather precarious when they are pressed beyond a certain point. The fact that temperate zone ecosystems are tough enough to survive a winter means that they can adapt to ruthless and drastic changes introduced by the human race. Tropical ecosystems always had it easy, and so the strains imposed by the human race may be too much for them. We have had examples in the past of tropical collapse, as in the Mayan and Khmer societies.

As far as the rich countries are concerned, the critical issue is, first, how serious may be the real price increases in energy and materials? And second, what is the probability of a slowdown in the rate of increase of knowledge and know-how? On the first point, a modest optimism seems in order. Energy, for instance, absorbs only about 7−8 percent of the United States' GNP, even after the repeated oil price increases decreed by OPEC. A model prepared for the National Academy of Sciences Committee on Nuclear and Alternative Energy Systems in 1977 suggests that even a fourfold rise in the price of energy by, say, 2010, given some moderately reasonable assumptions about elasticities of supply and demand, would not affect the GNP very much. Such a rise in price undoubtedly would result in very large conservation and also push toward new energy sources, so that over a generation both supply and demand might turn out to be more elastic than we now think. Even if energy goes to 15 or 20 percent of GNP, this still leaves 85 or 80 percent instead of 93 percent for other things, and relatively small improvements in productivity in the non-energy fields would offset quite a large increase in the price of energy. In the materials field, likewise, though the situation is more obscure, there is a high probability for technological substitution in the case of almost all materials, except perhaps phosphates and of course, water, which may turn out to be the ultimate limiting factor.

Knowledge increase is very unpredictable. There are two opposite forces at work. One is the principle that we get to know the easiest things first, so that as science progresses, it gets harder and harder to find out more. Counteracting this is the principle that the more we know, the easier it is to acquire more knowledge. This results from the nonadditive or interactive quality of knowledge. Whether the first or the second of these principles will dominate in the next hundred years, no one can say. There are distinct signs, however, of a slowdown in radical change, whether technological or social, and it is at least a hypothesis of reasonable probability that while the rich countries may continue to get somewhat richer, they will be getting richer at a much slower rate than they have done, for instance, in the last generation. For the poor countries a substantial increase in know-how seems perfectly possible given the appropriate social institutions, simply because they are at a low level now and have a great deal to learn. Even for the poor countries, energy and materials limitations may turn out to be less important in the long run than now seems plausible, although we do not know enough about the energy and materials requirements for periods of rapid growth from a low base. The critical problem may well be the old-fashioned one—the difficulties of accumulation at low income levels. This is why the fairly universal austerity that comes with greater equality may be a critical factor. The egalitarian societies like China may simply save more and accumulate more than those with a high-living upper class and a large population at the subsistence level where accumulation is impossible.

Altogether the uncertainties are very large. Let us not be deceived by the apparent simplicity of numbers. Trends in growth figures are particularly deceptive. The differences between the rich and the poor are qualitative and structural even more than they are quantitative. The rich and the poor represent different ecosystems of human knowledge and artifacts. Development into riches is not just "growth"; it is a profound change in the knowledge structures and behavior and skills of the population and a product mix both of outputs and of the stock. It is fair to say that the rich countries are rich and the poor countries poor today not primarily because of exploitation, which is a minor though real element in the process, but because the rich countries underwent a process of radical, socio-ecological transformation and the poor countries did not. It is almost certainly easier for the rich countries to sustain their present structure than it is for the poor countries to attain it. The eradication of poverty and the development of a world that is reasonably uniformly rich is an objective for the human race that commands our very deepest assent. But the fulfillment of it is difficult and may take a painfully long time. Impatience may speed up the process but can also easily set it back.

NOTES TO CHAPTER TEN

1. Kenneth E. Boulding, "Reflections on Poverty," in *The Social Welfare Forum, 1961* (New York: Columbia University Press, 1961), pp. 45-48. Reprinted in *Kenneth Boulding/Collected Papers*, ed. Fred Glahe (Boulder: Colorado Associated University Press, 1971), vol. 2, pp. 181-96.

2. P. Menozzi et al., "Synthetic Maps of Human Gene Frequencies in Europeans," *Science* 201 (September 1, 1978): 786-92.

About the Editors

Franklin A. Long is Henry Luce Professor of Science and Society at Cornell University. He is the author of articles on science and public affairs and on technology for development and is coeditor of a comprehensive study of *Arms, Defense Policy and Arms Control* (1975).

Alexandra Oleson is Assistant Executive Officer of the American Academy of Arts and Sciences. She is coeditor of *The Pursuit of Knowledge in the Early American Republic* (1976) and *The Organization of Knowledge in Modern America, 1870–1920* (1979).

About the Contributors

V.V. Bhatt is a development economist and currently a Division Chief in the Development Economics Department of the World Bank. He is the author of several books and articles, his most recent publication being *Some Aspects of Development Strategy and Policies* (1978).

Kenneth E. Boulding is Distinguished Professor of Economics at the University of Colorado, Boulder, and a program director in the University's Institute of Behavioral Science. He is the author of *Ecodynamics: A New Theory of Societal Evolution* (1978) and *Stable Peace* (1978).

Harvey Brooks is Professor of Technology and Public Policy at Harvard University, Chairman of the Commission on Sociotechnical Systems and Cochairman of the Committee on Nuclear and Alternative Energy Systems, National Academy of Sciences/National Research Council.

Hyung-Sup Choi is former Minister for Science and Technology, Republic of Korea and currently President of the Korea Science and Engineering Foundation.

Paul H. DeForest is Associate Professor of Political Science at Illinois Institute of Technology. He is the author of a number of articles on science, technology, and public policy and is a regular contributor to *The Bulletin of Atomic Scientists*.

Robert Dodoo, Jr., is Secretary, Science and Technology Planning and Analysis Group of the Council for Scientific and Industrial Research, Accra, Ghana.

John D. Montgomery is Professor of Public Administration at Harvard University. He is the author of *Technology and Civic Life* (1974) and coauthor of *Values and Development* (1976) and *Patterns of Policy* (1979).

Dwight H. Perkins is Professor of Modern China Studies and of Economics, and Chairman of the Department of Economics at Harvard University. He is author of *Agricultural Development in China, 1368–1968* (1969), editor of *China's Modern Economy in Historical Perspective* (1975) and coauthor and editor of *Rural Small-Scale Industry in the People's Republic of China* (1977).

Gustav Ranis is Professor of Economics at Yale University. He is coauthor of *Development of the Labor Surplus Economy* (1964), *Growth With Equity: The Taiwan Case* (1979) and the author and editor of many other books and articles in the field of development.

Langdon Winner is Associate Professor of Political Science and Technology Studies at the Massachusetts Institute of Technology. He is the author of *Autonomous Technology: Technics-out-of-Control as a Theme in Political Thought* (1977) and of numerous articles on technology and society.

Index